D0396621

VENOM
DOC

VENOM DOC

THE EDGIEST, DARKEST, STRANGEST NATURAL HISTORY MEMOIR EVER

BRYAN GRIEG FRY

Arcade Publishing • New York

First North American Edition 2016

First published in Australia in 2015 by Hachette Australia Pty. Ltd.

Arcade Publishing books may be purchased in bulk at special discounts for sales promotion, corporate gifts, fund-raising, or educational purposes. Special editions can also be created to specifications. For details, contact the Special Sales Department, Arcade Publishing, 307 West 36th Street, 11th Floor, New York, NY 10018 or arcade@skyhorsepublishing.com.

Arcade Publishing® is a registered trademark of Skyhorse Publishing, Inc.®, a Delaware corporation.

Visit our website at www.arcadepub.com.
Visit the author's website at www.venomdoc.com.

10 9 8 7 6 5 4 3 2 1

Library of Congress Cataloging-in-Publication Data

Names: Fry, Bryan (Bryan Grieg), author.
Title: Venom doc : the edgiest, darkest, strangest natural history memoir
   ever / Bryan Grieg Fry.
Description: First North American edition. | New York : Arcade Publishing,
   2016.
Identifiers: LCCN 2016020316 (print) | LCCN 2016031723 (ebook) | ISBN
   978-1-62872-699-2 (hardcover : alk. paper) | ISBN 978-1-62872-706-7 (ebook)
Subjects: LCSH: Venom—Research—Anecdotes. | Poisonous
   animals—Research—Anecdotes. | Fry, Bryan (Bryan Grieg)
Classification: LCC QP632.V46 F79 2016 (print) | LCC QP632.V46 (ebook) | DDC
   615.9/42--dc23
LC record available at https://lccn.loc.gov/2016020316

Cover design by Laura Klynstra
Cover photo courtesy of Bryan Grieg Fry

Printed in the United States of America

To my loving parents, Astri and Jim,
for nurturing my life.

To my darling wife, Kristina,
for awakening my soul.

CONTENTS

1
WELL-SPENT YOUTH

There is nothing quite like agonizing withdrawal from months smashed to the gills on an extremely potent opioid derivative to turn to ashes one's fire for recreational drugs. Intravenous hydromorphone had been my neurosurgeon's weapon of choice for the last four months—a necessary remedy after a spectacular lifelong, out-of-control, self-medicating, self-destructive vortex had culminated in me being stranded in the City of Angels with a broken back. What a wild ride it had been. The kind of adventures where one lives decades in only weeks. My obsession with venom had taken me across the globe to seek out the world's most dangerous animals in the world's most inhospitable places, including conflict zones. I had been in and out of hospitals throughout these adventures. While there had been bodies on the floor along the way, I was still alive. For now. There was, however, the small matter of my paralyzed legs.

Thirty-eight years earlier, I came to my first awareness of self with my head restrained and all limbs strapped to a bed.

Intravenous lines had been surgically implanted into my temples and on the insides of my ankles. In this earliest of memories, I was in hospital being pumped full of a wide variety of chemical combinations, all in a desperate attempt to cure me of the spinal meningitis that was wreaking havoc on my nerves. At only two years old, I was one very sick little baby. My spine was cold liquid fire and my newfound existence a tortured hell. The reason I was restrained was that I kept grabbing onto the tubes like a hairless little monkey and pulling them out, even the ones inserted into my temples.

Eventually the electrical storm passed and the cleanup began. This was my first flirtation with death. Well, it was a bit more than flirting; bodily fluids were definitely exchanged. It was a hell of an introduction to the comedy club known as being human.

I had started walking just before I fell ill, but left the hospital so weak that I couldn't stand. I was back to square one, learning to walk all over again. My paternal grandmother, Gene, bought me a big toy truck so I could brace myself standing against it while I took the uncertain steps to rebuild my wasted leg muscles. During the follow-up treatment at the Walter Reed National Military Medical Center in Washington, DC, the team of neurology specialists had an intricate set of exams planned to test the recovery of my neurological function. However, these were immediately suspended after I spontaneously started doing multiple somersaults across the floor. My cavorting was all the evidence they needed to confirm my successful recovery.

It turned out that in this first critical event I had escaped without any long-term damage except for the hearing in my right ear being almost entirely wiped out and my sense of balance permanently affected. I had perfect hearing in one very narrow range, but on either side of that it was like my hearing had just been deleted, leaving the ear useful for not much more than hanging sunglasses off. It did come with the social bonus of being able to put annoying people on that side, particularly during a movie or a long dinner party. Plus, I could tune the world out by sleeping on my left side: the ultimate noise-canceling headphone. I now had a daily reminder of the concept of mortality—a concept I became familiar with pretty much in sync with the concept of "me."

My parents became used to my cavalier attitude towards life and its normal constraints. It was quite evident early on that I was not a flower child, rather, I was a flower piglet. Once, when I was three and outside playing contentedly with worms in a mud puddle, my mother saw it and called out to me, "Get in here and get those filthy clothes off RIGHT NOW!" I obeyed to the letter, and a minute later she looked out the window again to see my bare little bottom sprinting back out and flopping back into the same mud puddle. At times I can be literal to a fault.

I was at my happiest wandering blithely through the woods, flipping logs and rocks. I quickly proved myself adept at catching whatever lay beneath. My parents quickly realized that this was more than an interest, more than a passion: it was a calling. While it was certainly not a career path that they might have imagined for me, they were nevertheless extremely supportive. My father was the consummate hunter and fisherman, and he always encouraged an interest in the outdoors. I was lucky to have parents who had their own atypical pasts and interests, and thus avoided the more typical reaction of parents when confronted by an oddball. Which, in most families, could have been along the lines of "The boy is deranged. We need to see a psychiatrist. I told you not to drink during pregnancy!"

Very early on, I was struck with a deep and abiding love for all the nasty little creatures out there. I was only four when I grandly announced I would study venomous snakes when I grew up. I was not into fluffy bunnies or downy ducklings. I liked my creatures sharp and full of chemicals. Just like me. My first proper envenomation came from a decent-sized bullhead catfish while we were living in Alabama, first from getting a pectoral fin spine into the meaty part of my thumb, and then a pectoral spine into the leg as I dropped the catfish in pain. I quickly got up to speed with just how agonizing defensive venoms can be and experienced the joys of the inevitable puncture wound infection over the coming days.

There was definitely a bit of the genius/crazy quicksilver shooting through the veins of the family tree on my mother's Norwegian side, which had spawned not only the epically bloodthirsty Vikings, but also luminaries such as my distant relative, the composer Edvard Grieg. I grew up listening to his music on the turntable, with "March of the Trolls" quite naturally my favorite piece. The unique wiring of my brain to begin with, and the frying of the circuit boards upon start-up, proved to be a potent nature-nurture combination. I had an ability to focus upon one thing to the absolute exclusion of all else, tunnel vision the likes of which my parents had never seen. Obsession and compulsion are not disorders: they are competitive advantages! Such a gift, however, was accompanied by the social skills of a stoned dingo.

Because of my father's career in the United States Army Corps of Engineers, we bounced around from military base to military base across the United States, moving almost every year. Our summers were spent back in Norway, my mother's homeland. This was where my brother was born and where I spent most of the months of my gestation. The constant travel afforded me the opportunity to keep feeding my appetite for new experiences and new animals; for me, the continual upheaval was a wonderful way to grow up. I thrived on the chaos. It resulted in a certain adaptability and, in a sense, universality—while language and culture change, people are still just people.

I was one of very few children not leaving the house unless I had my snakebite kit stuffed into a hip pocket of my cargo shorts. Razors at the ready to slice the flesh in case of a bite, and ammonium carbonate tablets on hand to sear the sinuses; we now know that the former is dangerous, causing great harm and no benefit, and the latter useless. In first grade, after moving to Florida for the second time, I had been in my new school for less than a week when I fell in love with the librarian. She had me at, "Would you like to see me feed a live mouse to my corn snake?" *Hell yeah, baby!*

At that time, Florida's natural world was not virgin by any stretch of the imagination, but it hadn't quite hit the truck-stop whore stage that it has now. Native animal species were still plentiful, but the first tide of alien species had begun washing up on the shores. The Cuban anole was a sign of the times: just another honest capitalist lizard fleeing communist Castro. The introduced marine toad ironically hated salt water but if there was an outdoor pool that was not screened, these toxic Jabbas could be found in the morning serenely floating on the surface. The pool screens were also my richest hunting grounds for basking lizards.

Snakes have a pair of glands located just inside their cloaca (the posterior opening that serves as the only opening for the intestinal, reproductive, and urinary tracts), about 80 percent of the way down their belly, which produce a noxious secretion. Some snakes have evolved these glands for active defensive use. Ring-neck snakes in particular, as it turned out. These snakes are glossy black on their backs, patternless except for the namesake narrow orange ring that circles their necks. The underside of the tail, however, is vibrantly colored with oranges and yellows. When they feel threatened, they curl the tail up tight like a corkscrew, displaying the color. Naturally, this did not deter me when I encountered my first one and I was sprayed with eye-watering stuff that was disgusting beyond description. Ordinary soap was summarily defeated by this foul concoction. I went around the rest of the day reeking of the delightful aroma of Ring-Neck Snake Funk #5, thus deterring even further my peer-level female schoolmates, who thought me rather odd to begin with, as I liked snakes and kept bringing them to school with me.

At this time we lived just a few blocks away from the iconic Miami Serpentarium, run by the legendary Bill Haast. Seeing him milk the king cobras left an indelible, golden-scaled impression in my soul. Black mambas moved silkily across their cages while rattlesnakes played a one-note sound of warning: a heavy-metal, open-E-string sort of droning buzz. One day my family was supposed to go to the Serpentarium again but went saltwater fishing instead. The king cobras would be there tomorrow. Or so we thought. We headed out into the sort of hard dawn that was characteristic of the Keys at times: harsh, unsubtle coloring with that trailer-park beauty queen kind of attractiveness. The two-stroke engine buzzed like a giant mosquito as we cut through water that looked like polished glass on this windless day. After a sporting time pulling in grouper and assorted smaller reef fish, we headed out into deeper waters than usual because of the calmness of the day. And there we struck green and gold: dolphinfish. The sleekest, fastest things on fins that I had ever encountered. My golf-ball-sized biceps were weak with exhaustion.

Back on land and starting our return home with an ice chest full of fresh fish fillets, we heard on the car radio about the tragedy that had occurred just moments before. Crocodiles at the Miami Serpentarium had killed a young boy after he had fallen into their pen. His father had irresponsibly let him stand on the ledge overlooking the pen. The boy turned and asked, "Daddy, what would happen if I fell in?" His father said, "I would save you." Ironically, this was not to be the case when the boy fell as he turned back to look at the crocs again. He was rapidly dismembered and partially consumed before Haast shot dead the two crocodiles. Tragically, the young boy and the two crocodiles all paid the ultimate price for the father's shocking level of parenting. The Miami Serpentarium closed immediately and that was the end of an era.

Two years later, as a northern alligator lizard relentlessly chewed on my classmate's finger like Luis Suárez having another mental breakdown on the soccer field, I noticed that the saliva coming out of its mouth was unusually frothy. But any mental notes about this were soon washed away by the copious amounts of blood that emanated from the wounds. This was all accompanied by the strange squeaking/squealing sounds the boy made every time the lizard clamped down and chewed. The only thing I was certain about was that it was entirely his fault; I had warned him about putting his finger too close, saying, "It will bite!" Which it certainly did, without hesitation.

Not unexpectedly, this commotion attracted the attention of the teachers. In the course of the post-mortem interrogation, it was revealed that during the previous six months at my new school on Hamilton Air Force Base in California, my mother had been dropping my brother and me off at the same time for school. The school had a weird staggered start, where grades 4–6 started at 8 a.m., while grades 1–3 started at 9. My brother was two grades ahead of me, so he started in the first wave. It made no sense for my mother to make two trips each morning and again in the evening, so I was simply dropped off with a free hour before school, while my brother had an hour to kill at the end of the day. As I was without explicit instruction regarding where to go or what to do (implied instructions being lost on my very literal brain), as soon as the car turned the corner, I would head straight for the nearby woods to go snake hunting. So each morning I would arrive back for class with my backpack writhing mysteriously with whatever the morning's hunt had turned up. This particular morning I had flipped a large rock and found the biggest northern alligator lizard that I had ever seen. The body, with its brown and white shingled scales, was flawless, all the way down to a perfectly intact tail. It was very rare to find a large adult without at least part of the tail regenerated after a potential predator was left with the rest impotently squirming in its jaws. I was made to let it go, which I protested loudly against since it would not change the damage to his flesh, which had required ten stitches to close the profusely bleeding wound.

When not causing administrative problems at school, I was invariably out snake hunting with various equally snake-mad friends. One of my favorite spots was a small lake flourishing with amphibians and snakes. I would hit it early in the day, the soft morning light painting the landscape with pastels. By 9 a.m., the temperature would already be in the mid-seventies—perfect for going around rocky areas and looking for cylindrical serpentine bodies halfway out of the grass and basking on rocks. The pond was littered with the multi-colored polka dots that were the lily pads and their flowers. Herons strode majestically through the shallows. Feral Florida bullfrogs provided a sustaining meal for indigenous wading birds.

One particular morning, we weren't the only ones looking at a body moistly glistening in the new day. Like the fingers of god, shafts of light came through the stone-blue clouds, revealing a "painting"—an unexpected nude portrait. For, standing in the middle of the clearing, striking a pose the likes of which I had never seen, and putting a finger somewhere my nine-year-old brain had never thought of sticking a finger, was the first naked woman I had ever seen. And she was being photographed by a mustachioed man with an oversized camera. The forest was silent as we all contemplated each other. A pair of scruffy urchins with mud-stained pillowcases containing moving creatures of uncertain identity. And them. We just trod on past, staring intently out of the corners of our eyes, experiencing odd new glandular sensations. While life went on, it certainly had a new flavor to it.

Not long after this came the championship game for my Little League Baseball team. As we were warming up, my attention was caught by an aerial traffic pattern of hornets going in and out of a subterranean nest located within a gopher's burrow. My fast but typically wildly inaccurate method of throwing objects was unusually on-target as I zinged dirt clods into the hornets' nest from a short distance away. I was momentarily distracted from tormenting the hornets by having to throw a ball back to the rest of the team, so I didn't notice until it was too late that an enraged stream of hornets was pouring from the hole. I copped fourteen stings in total on my arms, face, and neck. Of the many hornet stings I accumulated that day, the one that really hurt was the disfiguring one on my left eyelid that looked like a small tumor. It was off, again, to the Hamilton Air Force Base Hospital, where I was already on a first-name basis with some of the staff, and I missed out entirely on the championship game.

At this time, California was flush with funding and was wisely putting a considerable amount of this into education, including a flourishing Gifted Program, into which I was enrolled. In addition to learning Spanish and computers, we also competed in the regional science fair. My project, for which I won first prize, investigated whether my pet tarantulas were more likely to eat a prey item quickly if another tarantula was in view. After feeding my pets under a variety of conditions, I observed that in the presence of another spider they ate considerably faster; which I concluded was the arachnid equivalent of a kid stuffing his mouth with the last piece of chocolate birthday cake before someone else could eat it.

At the end of the third grade we went back to Norway for the summer. This was when I saw my first arctic viper, but tragically one that had been run over by a car not long before. A black stripe zigzagged down a dusky body. The tail was still reflexively wiggling from lingering nerve action and the oozing blood on the road had not yet clotted. While Norway has the most infernal winters, it has the most delightful summers. Never-ending days spent exploring the craggy rocks that my overactive imagination turned into the trolls of lore. It was also a rich playground for weird animals. One time, while at the family island home on the Asker fjord near Oslo, I asked for a bottle for a fish I had caught with my hands under the jetty. Eyebrows rose when I selected a long, skinny wine bottle. Naturally there was curiosity as to exactly what sort of fish I had caught. This would not be the first occasion I would target and capture an animal in another country that the locals didn't know existed. In this case, it was a banded pipefish, a relative of the seahorse. Other animals I successfully sought out during that trip included the Norwegian crested newt, which reminded me of the dragon Fáfnir from Norse mythology.

Upon our return to California, the house had a certain pungent odor to it. As it transpired, a family of skunks had moved into the crawl space at the back of our basement. My father dutifully notified Base authorities, who, try as they might, could not find the right path through the inflexible bureaucracy to remove them. Local animal control did not have authorization to come onto the military base until the proper paperwork was shuffled. Which, of course, would take days. There was no one in the military chain of command tasked with dealing with chemical-weapon-laden skunks. As my father was a full colonel by this time, the staff knew exactly who and what he was. The various ribbons and such on his uniform were code for acts worthy of recognition, some of which came during his two tours of Korea and one of Vietnam. So it was accepted by all when, with tacit approval, shots rang out from under our house. A full load each from a large-bore shotgun and a .45 pistol rendered the problem immobile. Various bodily fluids painted the wall. However, this resulted in a dramatic increase in the smell. The invisible chemicals hit us with the same intensity as that experienced by soldiers in World War I mustard-gas-filled trenches. The house was now temporarily unliveable. By the time my brother and I were done helping my father bag the carcasses and then bury them out in the woods, our clothes were suitably ruined and were summarily burnt. To us kids, this was one of the best days ever!

While my mother forbade me to run around barefoot, this lasted only so long as it took me to get out of her view, then off came the shoes. I liked the feeling of the dirt under my feet and the grass between my toes. The game was up, however, after one typical day spent wandering through a creek catching rough-skinned newts when I sliced the sole of my right foot deeply on a glass bottle that some cretin had thrown into the murky water. As it was being stitched up after yet another trip to the Base hospital, my mother naively exclaimed, "I don't know how this could have happened. He never goes without shoes!" The doctor gave me a knowing look and said, "Lady, this kid has feet tougher than my dog."

After five idyllic years in California we were off north to

Portland, Oregon. People didn't tan there, they rusted, and the state bird was a mosquito. The only saving grace was the stellar salmon and steelhead fishing. The offshore fishing at this time was also phenomenal. It wasn't long before I clicked on one little tidbit that seemed to have escaped the notice of others—there was a mighty discrepancy between the spearfishing records and line fishing records. So for my sixteenth birthday I requested an offshore fishing trip with my father. The sky was cold fusion blueblack as we met the boat and the skipper at the Garibaldi marina in Tillamook Bay at 5 a.m. I had a well-researched list in my pocket of ten species for which the spearfishing world record was dramatically different from the line fishing world record. These were all very common fish that were the bread and butter of the charter boat industry. It was just pure oversight because they were weekend-warrior sort of fish, not the sort featured on fishing shows.

The day started productively, getting plenty of decent-sized but far from full-grown black rockfish. Then it happened. I pulled up a massive cabezon, the largest species of sculpin. Huge, hand-like pectoral fins jutted out from the big-mouthed, toad-shaped body. It was not massive in the grand scheme of things, considering the spearfishing record was a bit over twenty-six pounds. However, crucially, the line fishing record was only nine pounds, as this was viewed as decent eating but a trash fish, not a sport fish like salmon or marlin. I took one look at this fish and started to get very, very excited. It was definitely larger than the record! The crew met this with much confused amusement. One of them drawled, "Boy . . . I've caught ones of that same type you could fit this one into the mouth of." To which I cheerfully responded, "Yeah, but did you report it to the International Game Fish Association?" which elicited the reply of, "Uh . . . the International Game what?"

Back at the marina, the fish topped the scales at nearly seventeen pounds—a new world record. Once we were done with all the measurements and certifications my father asked if I wanted it stuffed and mounted. I replied in true Viking-child manner, "Nah, it's too ugly. Let's just eat it instead." The blue-green flesh had a decidedly alien quality to it, but baked with bacon, chilli, garlic, butter, and lime it turned out to be extremely tasty. Easily the most satisfying fish meal I have ever had.

My world record would last less than twelve months, as long as it took for another annual round of entries. It turned out that despite the previous record having stood for over ten years, I was not the only one who had noticed this oversight. In a remarkable coincidence, someone that same year had in fact submitted an eleven-pound specimen, which I am sure they thought was a sure bet. This flurry of activity naturally caught the attention of others and the following year someone submitted an eighteen-pound specimen; within a few short years, the record was up to a much more realistic twenty-six pounds.

I had been maintaining, at this time and at home, a large collection of non-venomous snakes, with the kitchen freezers containing foil-wrapped dead mice to feed them. Once, while in a hurry, I tried to defrost a mouse in the microwave. With the incredible timing of mothers the world over, my mother walked by and glanced at the microwave at the exact moment the mouse swelled alarmingly and then exploded. The painting of the microwave walls with various internal organs was accompanied by an incredibly noxious smell. She was not amused.

Two weeks after high school graduation, I suffered my first snakebite. The animal responsible was a particularly beautiful timber rattlesnake, a captive-born baby. The parents were both caught near the Florida–Georgia border and were of the sunset pink type commonly referred to as canebrake rattlesnakes. I was over at my friend Richard Richey's place picking up the first venomous snakes that, after much pestering, I was to be allowed to keep at home. Small snakes can be much more difficult and dangerous to handle because the room for error is so small and, as I was examining the snake, one long, mobile fang flicked out of the side of its mouth and into my thumb. There was an almighty *KLONG* as the emotional wave of shit hit my heart and sweat ran down my brow. Over the next thirty minutes, there was a lack of the pain and swelling at the bite site that are the hallmarks of a rattlesnake envenomation. So, we came to the hopeful conclusion it was a dry bite and I proceeded to commence the drive home.

The first indication that maybe I hadn't got away scot-free happened ten minutes into the drive, when a strange metallic taste developed in my mouth. It wasn't long after that that something much more dramatic occurred. I lost my ability to see red, blue, green, or any color other than yellow. It was a monochromatic world ranging from white to black with only shades of yellow in between. "Uhhhhh, that's no good," I thought to myself. Reality retreated. Shapes swirled. Sounds distorted, some bursts echoing like the reverberating distortion for electric guitar often overused in eighties glam metal to cover a pathetic lack of skill (such as any of the offerings by the lipstick-wearing monstrosities in the band called Poison). I descended down the rabbit hole and into a hallucinatory vortex.

I managed to steer the car into a small service station that was the first place I came upon in this remote country area. I stumbled inside and asked the girl at the counter to please ring for an ambulance as a venomous snake had bitten me. A red-and-black-flannel-shirt-wearing, lumberjack-looking customer waiting asked, "Are you sure it was venomous?" My answer came in the form of me doing the "full *Exorcist:*" projectile vomiting all over the lottery tickets for sale, cash register, beef jerky, and all else in the front counter firing line. My eyes rolled up into my head and I collapsed into a convulsing heap on the black and white checker-patterned dusty linoleum floor.

During the course of the ambulance ride to the hospital, I suffered a series of heart attacks. It turned out that the venom of this particular population is unique in possessing neurotoxins of the sort more typically seen in rattlesnakes from Brazil. There were no local effects at all throughout the ordeal, other than the flesh wound from the fang. The neurotoxins, however, were of a decidedly sinister type and gave my nerves a great big chemical raping.

After much chaos and frantic activity in the emergency room, where I was definitely the star of the moment, things were stabilized and I was admitted to the Intensive Care Unit. I then had to ring the friend whose car I had borrowed. "Where's my car? I'm late for work!" was her opening gambit. My reply—that I was in the hospital and had no idea where her car was—was met with, "What the hell do you mean, you don't know where it is?!" I explained that I was hallucinating from snakebite and didn't know where I had ended up before an ambulance took me away. This was met with stunned silence. If she had been told I had been caught smuggling Mexicans back into Mexico, in search of a better life in a civilized country with a functioning medical system, she would have found that more reasonable and believable than what I had just said.

I was released the next day and my parents and I went to pick up my friend's car. It was quite amusing to watch my father, who cared for his own vehicles with military precision, driving this completely unroadworthy vehicle. It was of a type favored by female college students who view vehicle maintenance through the lens of "The oil doesn't need changing yet since the light hasn't come back on" and who parked using the "bumper car" method. Compounding matters was that it was a Renault, which when new would have been more intact but would still have demonstrated the shoddy quality of workmanship for which French union workers are notorious; a "construction" that is the inevitable result of an assembly line worker proving he is in control of his destiny by doing an artistically crappy job putting the car parts on. Several owners later in the hands of a *Legally Blonde* type, and voilà! One rolling wreck coming up. Notice that back left wheel wobbling alarmingly? It really shouldn't be doing that. Only the first and third gears worked; it handled like a Doberman on acid and cornered like a three-legged water buffalo.

Thus I passed into adulthood in my own inimitable manner and to Portland State University I went. Based upon the grades I obtained in high school and the strength of my SAT scores I had been awarded a scholarship that covered my tuition and most of my costs, so I only needed to come up with entertainment money. I didn't want to work during university terms so I set about becoming a bit more creative in making money. Working as a male model for print and runway paid well, but was very feast-and-famine. Working as a bouncer also brought in some cash. To further supplement my income, I settled upon selling my bodily fluids: plasma and sperm. Not in that order. Having plasma stripped out of the blood and the red blood cells returned was a weird sensation. One that also required a good meal afterward to replace the lost nutrients. Which in turn cut into the profit margin. But this procedure also left me in no shape for the selling of sperm to Portland's premier eugenics factory. So, it was always sperm first. I struck up a friendship with one of the nurses who worked there and she kept me up to date on how many times a "mini-me" was spawned. It ended up being nineteen. Selling sperm in a nicely decorated uptown medical facility with a quite amazing selection of high-quality porn was, of course, infinitely more pleasurable than joining the queue of winos outside the blood clinic, where I was always the stand-out in being the only one who had showered in recent history and had all my teeth intact.

It was in this period of time that I made what would become one of my all-time great friendships, a friendship that persists to this day. Arun Sharma looked like Gene Simmons would have looked if he were in an alternative universe as a motorcycle-riding Bollywood pimp living in Harlem during the late seventies. Long, tangled black hair set off a meticulously maintained goatee.

We met when I sold him a nicely patterned Asian water monitor. A week later he rang me and said, "Dude, this thing is possessed. I thought you said it was tame?" I was pretty confused because it was really chill when I had it. So I rolled up and opened the lid of the shallow, opaque-blue plastic box he had put it in after removing it from its cage ahead of my arrival. I saw the dark pink back of its throat as it rose straight up and deep-throated my thumb. It hung in the air suspended from my thumb, then started chewing vigorously. There was silence for a moment before Arun gave me the biggest, toothiest grin and said, with evident delight at this turn of events, "See? I fucking told you!" In order to get it off, we had to fill its mouth with clothes pegs with the ends pressed closed and then release them inside the mouth so that their springs would collectively fatigue the monitor's jaw muscles and allow my finger to be extracted. My finger emerged mostly intact, but over the next hour it became swollen and throbbed more than I would have expected from the mechanical damage alone; it also bled longer than I expected.

I swapped the lizard over for a gorgeous six-foot female red spitting cobra from Mozambique. Like all beautiful females, it was extremely high maintenance. Not because it was nasty—it was a totally relaxed animal that never spat once—but because it was in continual motion as it glided fluidly around the spacious cage. All that activity meant it was always hungry and required daily feeding of a pair of mice. This inevitably resulted in a daily crapping in the cage that this feral creature would then proceed to happily smear all over the cage and glass.

After this inauspicious beginning, Arun and I joined forces to start a reptile business, with much the same motivation as a drug addict who starts a drug business to support a habit. Running a reptile business with Arun was an amazing time. Not just because of the plethora of cool animals we obtained from all over the world, but also because of the intersection of people radiating from the core. There might be Desert Storm military types, home from a region of the world full of things far more dangerous than venomous snakes, checking out freshly hatched albino monocled cobras at the same time as some hippy-trippy types were gaga over rainbow-colored panther chameleons. People who would normally pass like ships in the night were instead on the same wavelength because of this broader common interest and would happily chat the hours away while beholding these pieces of living art.

Whenever we'd make a run to the airport it was like venomous Christmas. While we knew what we had ordered, what actually arrived might not strictly match the manifest, and in any case the shipments always contained something we'd never seen. In addition to overseas suppliers, we often got shipments from Florida, with Tom Crutchfield, Glades Herp, and Strictly Reptiles being the major connections. Clearing the consignments was usually pretty routine, with the airline officials not keen to inspect a shipment of reptiles, particularly ones containing venomous species. When we had to open a load for inspection it was done in a room off to the side and away from other customers. One time I was showing a shipment from Malaysia that contained cobras as well as blood pythons. The official was particularly keen to see the blood pythons; the name had caught his eye and sparked his interest as he had a pet boa constrictor at home. I was holding one of them when I was distracted by a question from him and it struck right for my crotch like a heat-seeking missile. Luckily, I was wearing sweat pants, so the long teeth got caught up in the fabric. The official's eyes bulged like that of a hypertensive toy poodle. It took me thirty very embarrassing minutes to extract myself from it and I left with my dignity as shredded as the front of my pants.

Going to university was a wonderful walk through the woods of knowledge. Because of my high-school grade point average and SAT scores, I was accepted into the university's honors program. This meant all the general university requirements—a collection of random courses that take up almost half of the time spent during the degree studies—were waived. While the intent of these courses was to broaden one's education beyond the narrow focus of one's chosen major, they were of supreme disinterest to me, so I was very happy with this arrangement. Instead, in addition to my molecular biology major, I was enrolled in a scientific philosophy co-major along with all other members of the honors program, leaving me with time to also do a psychology minor. I became fast friends with one of the biology department graduate students, Ed Degrauw, and he came in with Arun and me on several reptile shipments. We spent many happy hours having rambling conversations about all things reptile. Within my science major, there were not any courses on offer that dealt specifically with venom, so I packed my major full of courses on pathogenic microorganisms, learning all about bacteria, parasites, and viruses.

The lessons soon went from the abstract to the crushingly real. Towards the end of my undergraduate studies, it was evident that a good friend of mine's health had taken a turn for the worse. Jon had been diagnosed with HIV/AIDS when he was twenty, while stationed in Germany after joining the US Army straight out of high school. Everyone in his unit knew he was gay and no one cared. He was just an ordinary, average guy. The kind you could have a beer with while watching athletic men in tight pants chase a bouncing ball down the field, then slapping each other on the butt in celebration. Extremely intelligent with a delightfully twisted wit, he used to try to shock me. But come on, I'm me. I'm unshockable. Or so I thought. He did have a way with words when it came to describing geometric combinations that even if trapped on a desert island for a very long fucking time, such considerations never would have come to mind. Some, I reckon, I could have done without hearing. No amount of mental bleach could get rid of those stains. At the end of the day, however, I very much appreciated his dark elf sense of humor.

Once Jon was diagnosed, he was given an honorable discharge on medical grounds. He then promptly enrolled in the honors program, where we met on the first day. He was combining engineering with the required philosophy dual degree. I remember asking him one day why he was studying stuff he would never get to use. Why wasn't he lying on a sandy beach somewhere tropical with a high-octane drink in his hand? He replied that he wanted to see out his days doing what he loved the most: learning. He ended up having to drop out during the fourth and final year. As he became more skeletal, he became progressively weaker and tired very easily; he was in serious decline. His immune system was now totally wiped out—he was like a computer hooked up in a Thai Internet café with no antivirus software.

The summer before my last year of university, I went to the University of Southern California to do my first real stint as a venom researcher. The data I obtained later formed the nucleus of my molecular biology honors thesis. I had been awarded a travel grant to head down to the laboratory of Professor Frank Markland, one of the premier researchers on the medical use of snake toxins. My project was to test the efficacy of a peptide called contortrostatin, from the venom of the southern copperhead pit viper, in blocking the spread of Kaposi's sarcoma, the type of cancer that causes the characteristic skin lesions in HIV/AIDS patients. So it was with no small amount of bitter irony that, as Bang Tango's song "Someone Like You" blasted out of the lab stereo, I got the phone call about Jon's death. Ultimately, it was an ordinary bacterium that infected his blood and killed him.

2
ENDLESS SUMMER

On December 14, 1996, I opened an envelope from the University of Queensland with great anticipation. The fact that it was a thick one gave me hope that my application for a PhD scholarship to the Centre for Drug Design and Development ("3D Centre") had been successful. This turned out to be the case. Elation flowed through my veins like a shot from a drug. It was the purest joy I had ever felt. After jumping through the remaining hoops and obtaining my student visa, I departed on what turned out to be Australia's national holiday. I was confident this was an auspicious sign for my new life and long-sought-after dream to become a professional venom researcher.

After thirty hours of travel, I arrived a bit wrecked but extremely excited. My initial sight of Australia from the air was of the serpentine Brisbane River twisting its way through the city and entering Moreton Bay with a muddy plume that looked like a flat tornado.

The first thing I noticed was that the Australian peak summer was as hot and humid as a walk-in sauna. Not long after I arrived, it started raining so hard that it was like being hit with a fire hose set on full blast. I figured this would be good snake-catching weather, so I went out cruising that very evening with one of my new housemates. She suggested we hit nearby Mt. Glorious—a truly appropriate name for what turned out to be a stunning rainforest, filled with brightly colored parrots that zipped around from tree to tree like hyper young children, making more racket than a bingo room full of retirees. These were not the sounds of the songbirds of my youth; these were raucous screeches and screams more appropriate to a metal singer than the tricolored elegance flapping before me.

Nightfall was greeted with a crescendo of frog calls of all description that grew in volume until we had to yell to hear each other while wandering around a pond near Jollys Lookout. The first snake I saw was a golden crown snake, its iridescent brown body set off by the namesake colored head—a venomous species I, of course, had never seen. It is a member of the Elapidae snake family which meant it was characterized by hollow, short, stubby fangs set far forward that were linked to a muscular compression system capable of delivering a high-pressure stream of venom; basically the snake equivalent of having a pair of hypodermic syringes tucked away in the head. As I hadn't yet organized my scientific collecting permits, it was a "looky but no touchy" situation, which left me a bit frustrated, but I got over it once I saw six more that evening. They obviously weren't in short supply, so I would have ample time to catch and milk some later on. For now, I could just relax and familiarize myself with this gorgeous new venomous playground.

After the first night's excitement I could not wait to get started on the research at my new university. The next day I made my way over to the St. Lucia campus of the University of

Queensland. It was even more beautiful than the pictures had suggested. Bounded by the Brisbane River on three sides, the university lacked distracting traffic noise and hosted diverse wildlife on the native-vegetation-filled campus. Black-headed ibis strode elegantly along, looking like Egyptian hieroglyphics, while bright green water dragons with their paint splatter of white spots looked at me with calm indifference.

My first impression of the 3D Centre was that there were lab toys the likes of which I had never seen. I had no clue what they did, but I was looking forward to finding out how to put them through their paces. The air buzzed and hummed with the feverish activity of many active brains running on caffeine-fuelled hyperdrive. My PhD supervisor, Professor Paul Alewood, was brilliant and congenial. He had a peculiar Jabba the Hut laugh that came out whenever he told an off-color joke, which was just about every other sentence.

We spent the afternoon mapping out the plan of attack for my PhD, the distillation of which could be written on the inside of a matchbook with a crayon: catch a bunch of weird elapid snakes and see what is in their venom. A broad brief that suited me just fine since it gave me a licence to play. I spent that afternoon painfully and patiently filling out the wildlife research permits, which were dreadfully organized and written in impenetrable English liberally spiced with bureaucratic weasel words.

Continuing my exploration of the campus I discovered that it hosted an Olympic-sized swimming pool among various other excellent athletic facilities. This confirmed the stereotype of Australians being sport fanatics. I also eyed off the beach volleyball courts before heading off to the athletics office to enquire about getting onto a swim squad and a beach volleyball team. Both objectives were sorted in short order.

I soon settled into a rhythm of biking to the lab at 4:30 a.m., when the sky was the color of an old bruise. I would take advantage of the early dawn that was the result of Queensland stubbornly refusing to change the clocks to daylight saving time during the summer. I would be showered and hard at work by about 5:15 a.m., giving me nearly four hours before most other people came in. Not only did I count each hour alone as two hours' worth of productivity, but it also gave me the opportunity to quietly clean up my messy mistakes or crack out a small screwdriver to make some emergency repairs on a vital piece of equipment before anyone could take note. I would then chat with the "late arrivals" over coffee for a half hour before heading off for swim training, lunch, and then volleyball training. I would return to the lab around 4 p.m. to grab a few more quiet hours once the hustle and bustle had died down as the others left at civilized hours. If I was really excited about something, it was not uncommon for me to glance up at the clock to realize it was already past midnight; more than once I left as dawn was suggestively flirting with the darkness.

The combination of extreme ultraviolet light, intellectual stimulation, and lots of sports training had my body firing along like a well-tuned machine. I was happy, energetic, and enthusiastic. I got tired of washing sand out of my hair and having it perpetually wet from swim training, so I decided one day to shave it all off, first with clippers, then a razor. Walking out into the rain that first time was one of the most erotic sensations I've experienced. The warm water impacted on nerves that had never been caressed in such a sensual manner. I liked the look and feel so much that I decided to keep it permanently.

Once I had successfully navigated the byzantine maze of the Queensland Parks and Wildlife permit system, I was armed with the snake catcher's equivalent of being a 007 agent: I was licensed to catch, keep, and use. I set about rampaging through the forests and deserts of Queensland. I could be found out cruising Mt. Glorious most nights, due to its proximity to campus and it being a biodiversity hotspot, either alone or with my mate David Quigley in whatever cheap, beat-up wreck of a car I was driving at the time, or in his much more suitable Rocky four-wheel drive. I was buying cars for a few hundred dollars, doing enough tinkering on them to get them to run, and then going out snake catching in a vehicle totally unsuited for anything outside of a parking lot, let alone steep mountain roads. One night while driving a Mitsubishi Sigma (or the Stigma, as I referred to it), its brakes gave out completely. The pedal was about as useful as male nipples. As it was an automatic, this posed special difficulties. I rammed it into the lowest of the semi-automatic options and pneumatically yanked up the handbrake when I needed to marginally slow down. Somehow I managed to make it home. As it wasn't worth fixing, I flipped a few hundred more to another dodgy car dealer and had yet another unroadworthy piece of crap that would last about as long as my typical romantic relationship, which was less than the lifespan of a large head of garlic in my refrigerator.

Like a kid in a candy store, it didn't take me too long to get a cavity; I copped my first snakebite two months after arrival. All snakes have characteristic movement patterns based upon their morphology. But the only way to predict how a new type of snake is going to move is to spend time working with one to get familiar with its particular proclivities. This was a luxury I did not have, since most of the snakes I was targeting were only very rarely kept in captivity, if at all.

One moonless Tuesday night I was out with a few mates at what looked to be a particularly good time for snake hunting. It had been really hot during the day and then a light shower came through. Not enough to cool things down, just enough to get the frogs moving at dusk, which soon brought out the predators. We had seen fourteen snakes already, including the venomous species I was after; these included golden crown snakes, rough-scaled snakes and bandy-bandy snakes. The bandy-bandys were a delight to behold as these gentle, docile snakes did an elaborate raising of the mid-body to present their strongly contrasting black and white rings. Somehow this was supposed to be threatening; I just found it endearing. As the iconic Australian song "Great Southern Land" by Icehouse played on the car stereo, the next snake on the road had none of the clean, crisp, cool coloring of the bandy-bandy. Instead, this snake was a muddy dark brown alternating with dirty cream rings. Only one snake in my book fit this description: Stephens' banded snake.

The book at hand was conspicuously lacking in detail about venom composition, which, of course, was why this species was a prime target for my PhD research. The lack of information became more significant once I was bitten. The snake had moved left when I expected it to go straight—a mistake on my part that led to twin drops of blood on my right index finger. The venom's effects were rapid and severe. I instantly had a pounding headache accompanied by a crushing feeling in my chest, like a giant was sitting on it. The world turned and the ground rushed up to meet me. I thought to myself, "Hmm . . . this is a bit different. Haven't read about a reaction like this to any snakebite. If I survive this, it would make an excellent PhD topic." Then, just like a warrior from *The Iliad*, darkness veiled my eyes.

I was out cold for ten minutes and then I was awake as if a switch had been flipped. I noted with alarm that the two puncture wounds in my thumb had not stopped bleeding. Not a good sign. In order to slow the flow of venom as it was absorbed into the lymphatic system ahead of entering the blood stream through a lymph node in my armpit, we applied a compression bandage and immobilized my arm in a sling. This was followed by a mad dash to the nearest hospital.

The first blood test was taken less than an hour after the bite had occurred, but the results indicated that my blood chemistry was completely disrupted. In the blood of a prey item, the venom would be more concentrated and result in several massive blood clots, causing the prey animal to quickly die of a catastrophic stroke. But in my larger blood volume, the venom had been diluted and instead produced millions of useless tiny blood clots called microthrombi. By themselves the microthrombi were not harmful, but the net result was that I had no raw materials left to make a blood clot if I really needed to. So, I was at great risk of bleeding to death. The venom had consumed all my clotting factors and my blood was like water.

Potentially lethal changes were also occurring to my heart rate and blood pressure. Despite my anxiety and fear, my heart was only beating forty-two times a minute. My blood pressure was 78/26, so low that I was at risk of multiple organ failure occurring at any moment. This freaked me out a bit and then, when I saw that my freaking out did not raise my blood pressure or heart rate, that freaked me out even more. There was something in the venom that was stabilizing me at these extremely low levels. This was a continuation of whatever had rendered me unconscious within two minutes of the bite. I wondered how low these vital indicators had dropped when I was unconscious, since I was now conscious with these extraordinary low values. To distract myself, I thought about how this might benefit a snake when feeding and concluded that it was an excellent way to immobilize a prey animal—knocking it out, thus giving the blood toxins time to form the killer blood clots. Basically, the mice would stroke out while knocked out.

While the resulting effect on my blood pressure and heart rate was potentially lethal, it was manageable with stimulatory drugs. However, I was now at the same risk of bleeding to death as a hemophiliac. A dark tide was slowly spreading under my skin where intravenous needles had been inserted in each arm, as the blood leaked out of me. Bleeding out of one nostril and then the other followed not long after. Then out of each eye, so that it looked like I was crying tears of blood. I could have rung up the lab and said, "I won't be coming in today—my stigmata are acting up." This was terrifying in a rather cool way. But once I started bleeding out of my anus, life was decidedly not cool. No matter what the cause, anal bleeding is never cool.

Since this was such a rare snake, and the few bites that had occurred were extremely poorly documented, it meant that the doctors were unaware of the best course of treatment. The snake venom detection kit (SVDK) laboratory test gave a strong positive reaction for tiger snake. This did not mean that I had been bitten by a tiger snake; rather, it was suggesting that the tiger-snake-specific monovalent antivenom might cross-react with the liquid death that was now coursing through my veins. So, we gave it a go with an ampoule. After waiting three hours, retesting showed no effect and so another ampoule was given. By this time, my girlfriend-of-the-moment had curled up in the bed with me to give me comfort. Around 2 a.m. she rolled over, tearing out both IV lines. The jagged holes left behind now started to gush blood at a steady rate. The thick gauze pads placed over them were soaked with blood within minutes. After a while the nurses worked out the metrics of how long it took each of the bleeds to go through a particular thickness of gauze, thus allowing them to balance the dressings on the different bleeds, so that they reached the changing point at the same time. Doing math like this also helped keep my mind off the horror show of the situation I had got myself into.

After waiting another three hours, the retesting again showed no effect, so yet another ampoule was given. By this point I was about to lose my mind from the stress of it all. This was a skull-fuck like no other, knowing that any wound, however tiny, over a long enough period of time could become dangerous. But it was the idea of bleeding into the brain and ending up a vegetable that filled me with the coldest of terrors. By the time three more hours had passed and it was time to retest, I was seriously considering other career options. Studying the flight patterns of some butter-fly, perhaps, or something else equally innocuous. But this time, retesting showed a slight effect and so another ampoule was given to speed my recovery. I was in the clear as long as I could keep from having my tubes torn out again or doing anything else that would promote bleeding.

Thirty hours after arrival, I was released from the hospital once my coagulation profiles reached 80 percent of normal and were continuing to rise. After a very long sleep I continued recuperating with some quiet time in the hammock, while being bathed by the warm afternoon sun. However, in the evening I cut my foot on a piece of glass when I walked into the kitchen. The orphan ringtail possum I was raising had knocked a vodka bottle off the fridge top. After cleaning it up, I settled on the couch to watch some cricket. About two hours later I glanced down and noticed a pool of blood on the tile floor. And that none of it had clotted. I immediately went back to the hospital, where a new battery of tests confirmed that my blood was again unable to clot. We concluded that some venom had been trapped in tissue somewhere and then worked its way into circulation once I was out of bed and more active. It was another ampoule of antivenom for me, and more hours waiting on tests and retests. Eventually my blood came back good and I was sent home again.

For once, I did not blaze around at full speed but had a very quiet, contemplative week. I was scared out of my mind by what had transpired. On the other hand, the less emotional and more objective part of my brain viewed all of this with fascination. From a research point of view, this bite had revealed very novel venom effects upon the blood pressure. I had inadvertently become my own breakthrough. After milking a few more Stephens' banded snakes, I discovered that the component responsible was a modified cardiac hormone that is normally used to regulate blood pressure, with it being released in particularly high levels during heart attacks. This hormone class, called natriuretic peptides, relaxes vascular smooth muscle such as that surrounding the aorta, the main artery leaving the heart, thus causing a beneficial drop in blood pressure. So it was a case of the snakes recruiting something normally useful and turning it into a weapon. My blood destruction had been caused by a mutated form of the ordinary blood-clotting enzyme Factor X. Only it was a thousand times more active than usual.

It turned out that many Australian venomous snakes have these blood pressure toxins in their venoms as part of a chemical arsenal. The Stephens' banded snake simply has it in extremely high amounts. However, larger snakes, such as taipans, also have natriuretic peptides in their venom, so I focused my efforts on taipans for the simple reason that these snakes gave huge amounts of venom and were also plentiful in captivity. Even though the natriuretic peptides were in lower amounts than in the Stephens' banded snake, this was more than offset by the larger amounts of venom I could accumulate.

Taipans are best described as ten-foot-long, copper-colored ballistic missiles with inch-long fangs at the end. These iconic snakes are the largest and most infamous of all venomous snakes in Australia, as they have the most dangerous bite of them all. Over the coming months, I successfully milked fifty taipans for their venom. Most of these were captive snakes, which meant that they had zero fear of humans. The implications of a bite were horrifying. In addition to causing bleeding issues even more problematic than that of the Stephens' banded snakes, their venom also has an extremely potent effect on the nerves, muscles, cardiac system, and pretty much anything else reachable by the bloodstream. The lethal dose of coastal taipan venom for a human is estimated to be 3 milligrams, but venom yields regularly exceed 100 milligrams, and one massive male specimen gave over 700 milligrams in a single milking. The straight fangs grow so long they actually start wearing holes all the way through the bottom jaw from pressing into the floor of the mouth every time the snake closes its mouth. Perhaps this is why they are always so cranky! I milked one by simply pressing the lower jaw on to the milking container with the snake's mouth closed. The fangs popped out the bottom and liberally ejaculated venom.

Milking these snakes was quite the challenge due to them being lean, strong, and extremely agile. When I would milk a person's pet snake, they would usually stand way back and watch the show. Wise of them, but of no help to me. These milkings typically occurred in cramped sheds or rooms packed with all sorts of clutter—hardly the ideal scenario when dealing with such psychotic serpents. I had one really close call when one of the snakes launched out of the cage immediately after I opened the lid and left a long scratch along my thumbnail. A fraction of an inch in any direction and I would have been in real trouble. These incredibly intelligent snakes left me a nervous wreck by the time I was done. But I was able to accumulate twenty grams of coastal taipan venom and ten grams of the even more potent inland taipan venom. More than enough to complete my PhD . . . or to wipe out every single person on campus. I opted for the PhD completion outcome.

About six months after I arrived, my supervisor Paul, as well as Peter Andrews (the director of the 3D Centre) and other senior staff, were lobbying the government for money to fund the Institute for Molecular Bioscience. As part of the schmoozing, they were showing Peter McGauran, the then Federal Minister for Science, around the lab. He was to meet with one of the other PhD students, who was to show him the collection of live funnel-web spiders, the lethal arachnids best known for biting very rich people along the North Shore region of Sydney. I found these spiders to be pretty cool in their own multi-armed alien sort of way. They are so eager to bite that they throw back their legs, arch their glossy black bodies and display their long fangs, each with a drop of venom at the tip from a bit of premature ejacuvenomation. I politely waited until after Minister McGauran had dutifully used a pipette to suck up a small drop of venom from one of the fangs before I called out from the other side of the animal room, "Check this out!" and slid out a six-foot-long indigo colored spotted black snake from its cage. I then asked, "Want to milk it with me?"

As I pinned the snake's head and got a good grip, Peter Andrews was shooting me death stares that promised my certain doom and deportation should anything go pear-shaped. Minister McGauran did very well with milking his first snake until the very end. The excitement, fear, adrenaline, and a strong snake combined into a perfect storm, and his hand started to shake. Before I could react he had dropped the snake, with its head landing on the crotch of his pants. There was no way I was going to suck the venom from there, so I quickly but lightly grabbed the snake mid-body and teleported it back into its container. All ended well. Minister McGauran showed the venom-filled container off at a political meeting that evening and I wasn't deported. Word came months later that the funding application was successful and that the Institute for Molecular Bioscience was going to be a reality. While the snake venom milking may not have been the decisive factor, or even a contributing element, I am sure that killing the Federal Minister for Science would not only have destroyed my budding academic career, but would not have done the funding application any favors.

A week after this I was bitten by a close relative of the spotted black snake, the Butler's snake. I was force-feeding a juvenile that was a fussy eater by gently pushing a euthanized pinkie mouse (a newborn mouse still devoid of hair) down the snake's throat using the plunger from a 1 cc syringe. All was going well until the snake decided it didn't want to deep-throat a baby mouse, especially one that it had just met. So, it used its throat muscles to try to expel the mouse at the same time as I started a fresh push. The combined pressure resulted in the plunger penetrating through the soft body of the mouse and out the other side, continuing harmlessly down the snake's throat. Harmless to the snake, but not to me, since my finger also went down the snake's throat, resulting in a certain familiar feeling as both fangs pierced my flesh.

While this species was very rare and I was conducting the first analyses of the venom, bites from most species of the same genus were only considered dangerous if the snake was an adult. This was because these snakes had venom that acted relatively weakly on mammals since they mostly fed on reptiles and amphibians, to which the venom was dramatically more toxic. My urine darkened a bit, which was suggestive of some muscle breakdown, and lab tests confirmed this, but the effects did not become severe enough to necessitate using up more of the extremely expensive antivenom. My muscles were rather sore the next day and I tired easily for the next two weeks.

Other than that, the bite was seemingly uneventful—until I noticed something strange. Over the coming days my sense of smell steadily decreased until it was completely gone. While this made snake cage cleaning less revolting and getting DNA samples from road-killed animals less nauseating, it did impact on other areas, such as not knowing if I stank from perspiring in Brisbane's sweltering humidity. It also made food much less fun than it had been. Eating delicate French cooking was like chewing on recycled cardboard. Only chilli-laden Asian cooking made an impact. My sense of smell started returning over the coming months, but only came back to about 50 percent of pre-bite sensitivity. It also came back warped. Some female perfumes smelled like chemical cleaning products instead of delicate bouquets of aromatic sensuality; some foods smelled like they were dipped in formalin.

My nose soon took another beating. In addition to being on a beach volleyball team and a swim squad, I was also boxing. While I had escaped injury in the previous years of boxing and bouncing, I finally got a significant one. During one sparring session, a gloved right hand snuck snake-like past my left hand defense and smashed my nose with a sickening wet crunch. I sprinted for the locker room to see my nose pressed over to the right side of my face like a Picasso cubist painting. I was a hunchback-level freak show. My first thought was "I'm never going to get laid again," which gave me the guts to manually wrench it back into position. I pressed home the top part with an audible *CLICK*! The world dissolved into black cotton candy as I almost passed out from the pain. But the blood was no longer coming out in red contrails, which was a good sign. The next morning I looked like a raccoon from the twin black eyes I was sporting. There was a tiny and permanent inward curvature of the left side of my nose that only my infinite vanity would notice. I decided, however, that boxing was done with me. I would concentrate my athletic endeavours on beach volleyball and swimming; I'd risk tendonitis but not mutilation with those two sports.

Working with such huge quantities of venom in the lab created an unforeseen problem, one with dire long-term implications for my ability to work with venomous snakes. I noticed over time that I would sneeze more frequently and with increasing force whenever I opened a container of dry venom and small particles of venom wafted up into the air currents like dust from a room long unswept. I didn't think anything of this, since venom proteins are too large to be absorbed into the body. This is why the toothless idiots at carnivals can milk a rattlesnake and drink the venom to show their god-like powers. If these morons had any gum bleeding, the venom would be able to enter their body through those wounds. Otherwise, they could take a bath in it without any ill effects. But I didn't consider the reaction of my immune system. Over time I developed a severe allergy to the venom. In the lab, this would manifest itself as symptoms similar to those experienced by people with allergies to pollen. It meant working either in the fume hood or wearing eye goggles and a sealed rubber mask with a particulate filter of the same sort I would use while laying down fresh fiberglass on a surfboard.

But these symptoms rang a strident warning bell, because it meant that if I was bitten again, I would almost certainly go into allergic shock from the venom, in the same way that someone who is allergic to bees would react to a sting, or a suitably sensitized child to a peanut. I was not yet done with the research I had set out to do, so I conducted a cost-benefit analysis and decided to keep on going. However, I ensured that I always had injectable adrenalin with me to counteract any shock, and also injectable antihistamine to hopefully prevent shock from starting back up again—a much higher tech version of the snakebite kit I had always had with me as a kid in Florida.

By this time I possessed quite the snake collection, including several specimens of the rare and very beautiful Pilbara death adder, with its alternating burgundy and black banding. One large female was particularly stunning and was easily my favorite snake in the collection. I didn't even milk her for venom. She was the only snake in the collection that was a true pet. She was fed first, cleaned first, and had the best cage set-up of them all. She would lie immobile in the cage until a defrosted mouse was presented, at which time she would strike with a speed I had never seen in any other snake. It was like quantum physics—she went from point A to point B without existing in between. But if there was no food around, she would not move for days on end. She would not react when my hands, heavily gloved, grasped the water bowl to take it out for cleaning. Because she was so calm, over time, without conscious thought I began to take liberties with her, such as reaching in barehanded to get the water bowl out. This worked fine for a very long time until one day she must have decided my fingers looked like something juicy to eat and she struck: she hit the back of my hand straight on. Rather than retracting, she used her long, mobile fangs to walk across the back of my hand, leaving a trail of six puncture wounds in her wake.

I immediately knew I was in mortal danger. Not just from the venom but also more immediately from the profound allergic shock I could feel coming on. My skin erupted into countless large hives, turning me into one giant itch. I felt like flaying my skin but then I was distracted by trying to vomit and breathe at the same time. My esophagus was being constricted to an ever-decreasing diameter due to the massive amount of fluid that was rapidly accumulating around my neck. This was a result of all the very small blood vessels internally leaking the fluid that transports the red blood cells. My blood pressure crashed like a stone as I desperately cracked open a glass vial of adrenalin, sucked it up into a needle, stabbed myself deep in the shoulder and rapidly pressed the plunger all the way down. This was quickly followed by an ampoule of antihistamine delivered the same way. The adrenalin raised my blood pressure enough that I was able to call out for one of my housemates to give me a ride to the hospital. I was living out in the bush on five acres in an area called Anstead, which meant that by the time the ambulance got there, we could already be well on our way to hospital.

I was shitting myself on the way to the hospital. Not just in the figurative sense, but also literally. My body was desperately trying to expel the allergic death-protein through every orifice. I spewed all over the dashboard, even down into the vents, and left a coffee-colored stain on the seat. It is hard to apologize with vomit bubbles coming out of both nostrils but I gave it a go. Halfway there I could feel myself going into shock again, this time accompanied by an alarming swelling of my face and shaven head. I cracked open another vial of adrenalin, pulled my black T-shirt sleeve up with my teeth and stabbed myself again. I glanced out the car window to see a suburban mother driving a station wagon staring at me in horror as she drove off the road and fishtailed on the gravel. Oh, what she must have thought of this pumpkin-head monstrosity shooting up in a car in the sterile suburb of Kenmore.

Arrival at the hospital was an anticlimax as far as the allergic shock was concerned, as the second shot of antihistamine had effectively prevented my mast cells from uncontrollably releasing more histamine. Block the release of histamine and everything else takes care of itself. Everything, that is, except the venom that was still coursing through my veins. By now it was starting to kick in. Initially, I was just a bit dizzy and uncoordinated. Which was pretty much my default state anyway, being naturally blond, having bad balance from childhood spinal meningitis, and being easily distracted by passing squirrels. But these effects became more and more severe until breathing became a struggle. My diaphragm muscle was being paralyzed by the neurotoxins.

As I became progressively more paralyzed and my breathing became more labored and inefficient, the most delicious sensation crept over me, like a technicolored chemical cloud. Blue gave way to black; my pupils became dilated and fixed. The lights became very bright and the colors very vivid in a way quite like being on psychedelic mushrooms. I was unable to open my eyelids or move my eyes, so my vision was limited to the times the doctor manually opened my eyelids to look at the pupils. The medical staff had no idea I was conscious and could hear everything they were saying. I just had no way of letting them know I was in there.

But I didn't care. The neurotoxins were now having an extremely potent narcotic effect. Life was beautiful. It was like breathing the most potent dental gas, times a thousand. Once I lost my ability to move at all and was put on artificial respiration, the sensation kicked up another gear and I was floating high above the world without a single care. True, I was locked inside my body, completely cut off from the outside world—the most primordial of fears. Strangely, I did not mind. This was entirely to do with the fact that I was having the most amazing party-for-one inside my immobile shell. Time warped. For aeons I drifted contentedly through the universe, exploring far-off lands and distant galaxies. This was a classic dissociative out-of-body-experience; a psychedelic state of mind that is reached by disconnecting the mind from the body, either by dissociative drugs like ketamine or, as it turns out, the neurotoxicity of certain toxins. Unlike a bad mushroom trip, however, I did not wonder if it would ever end.

Fortunately and unfortunately at the same time, the antivenom did its job and my Rastafarian world faded all too soon back to the much more mundane reality. The days, months, years, and centuries I had been traveling turned out to be contained within the eight hours I was fully paralyzed: a most interesting form of time travel.

This event was an excellent example of the wisdom that "it is the calm snake in the collection you have to watch out for." One never relaxes around a taipan or a black mamba. But, it is very easy to relax around venomous landmines like Gaboon vipers or death adders. This was combined with the supreme arrogance of a highly testosteroned twenty-something male. The sort that end up as hood ornaments on the karma of life if they were into motorcycles instead of venomous snakes. It makes perfect biological sense that the risk-assessment side of the brain of human males does not start developing fully until testosterone levels start dropping in a male once he reaches his thirties, and continues down from there with variable degrees of drop between individuals. Along with a calming of behavior. But back in the late teens and early twenties, this part of the brain is still in quite an embryonic state because it is okay if most of the males are causing carnage along the way, as each one can impregnate multiple females, such as in a pride of lions. We are still strongly influenced by our inherent animalistic nature, and this includes casual disregard of potential consequences.

Not long after this I heard about a very unusual death adder envenomation that happened to a person in Melbourne, where, in addition to the usual paralysis, the patient had severe damage to his muscles. I tracked down the patient details and gave him a ring. His name was Chris Hay and he had been bitten by his pet death adder, which was of a type that lives exclusively in the large floodplains of the black soil region in the Northern Territory area of Kakadu. He described his breathing getting shallower and shallower with each breath, until he gasped for a breath that simply wasn't there. Then the blackness set in and he faded away from reality. He could faintly hear voices, but they seemed so far away. While it was a terrifying and very lonely experience, it was strangely relaxing and the clarity in his mind was amazing. Upon awakening, he was shocked to hear three days had passed.

He sent me some of the venom and I conducted preliminary tests. Sure enough, the laboratory findings replicated the clinical effects. This was significant, since such potent action on the muscles was not a feature previously attributed to death adder venoms. Like my experience with the Stephens' banded snake, it showed that even catastrophic events like snakebites could have beneficial outcomes if all details were correctly documented. The venom caused permanent damage to his kidneys, such that if he became very physically run-down, the occasional squirt of blood would come out when he urinated.

By this point, I had amassed a sufficiently large pile of data, and a thesis of suitable quality could be carved out of it. The testing of the taipan natriuretic peptides had revealed that, consistent with the clinical effects I had experienced, they were more potent and longer-lasting than the ancestral cardiac hormones used to regulate the blood pressure. In addition, some subtle changes in their structure had also guided specificity toward two different receptors. Combined with their very small size, this meant that they had tremendous therapeutic potential. So it was with great satisfaction that we patented them for use in the treatment of congestive heart failure.

This was yet another entry in the long list of therapeutic uses of toxins, the stand-out of which has been the development of the high blood pressure drug Captopril from the venom of the lancehead viper from Brazil. This drug and its derivatives have an annual market of ten billion dollars. It reinforces the value of conservation, for if the habitats are wiped out, the animals will be extinct before we can study them. When people ask me what the best argument is to convince people of the value of conservation, I say that their weakest argument is to talk about how magnificent and wonderful the animals are. The only people who will appreciate that will be the ones who already think that way—it's very much a case of preaching to the choir. Rather, they should stress the value of conservation through commercialization, pointing out that destroying a stand of forest is no different than nuking mineral deposits. There is no way to predict where the next wonder drug will come from, so we need to conserve all of nature.

While hard at work writing, I took time off to go to a venoms conference that the 3D Centre was hosting on Heron Island. To get there required a boat ride across some very rough seas. On the two-level ferry, I took up residence on the couch inside at the front of the lower level so that I could get a nice view of the approaching oval, sandy island, without getting wet from salt spray. It also had the pleasant bonus of me not being one of the people coated when someone on the top level puked and the in-sweeping wind carried it all over the people below. It is bad enough to be covered with vomit, but even worse when it is someone else's!

I arrived on the island clean and dry and instantly decided to get wet and dirty. The rising tide was carrying with it extremely large stingrays who glided up from the depths to search the shallow reef flat for crustaceans to feed upon. Sliding otter-like into the warm azure water with my snorkeling gear on, I decided this was not a bad way to end the three and a half eventful years of my PhD studies.

3
SERPENTS OF THE SEA

By this time, my scientific outreach activities had caught the attention of documentary filmmakers. Not long after submitting my PhD thesis, I went to Townsville to film a BBC/Discovery Channel co-production called *Menacing Waters*. The basic premise was to show that, in one way or another, almost everything in the reef was using chemicals to kill or defend. The focus of my segment was the research I had commenced on sea snake venom evolution.

Many myths abound about sea snakes but the most persistent are that, one, they can only envenomate if they bite the person on the webbing between the fingers; and two, that they are the most toxic snakes on earth. Both myths are, like most myths, wrong. Sea snakes can envenomate quite readily, as many have fangs comparable in length to those of Australian land snakes. And while highly toxic, they are not inordinately so when compared to Australian land snakes. The comparison to Australian land snakes is appropriate, as sea snakes are actually their descendants and remain their closest relatives. A tiger snake is more closely related to an olive sea snake than it is to a cobra. This close relationship is also evident in the composition of venoms. There are two land snakes, the inland taipan and the eastern brown snake, that are more toxic than any sea snake, but from then on the sea snakes share the rankings alongside snakes such as tiger snakes and death adders.

Sea snakes are supremely adapted to life in the sea. We did deep dives with them but had to pull up at 115 feet. The sea snakes continued far down into the gloom, diving effortlessly to below 330 feet. These animals have an amazing ability to lose almost all their excess carbon dioxide and nitrogen through the skin, while simultaneously taking up an additional 20 percent of oxygen. This loss of carbon dioxide and uptake of oxygen means that the snakes have dive times of over an hour when active, and can remain submerged for three or more hours when sleeping.

The loss of nitrogen means that it is physically impossible for them to get "bent:" "the bends" is the medical condition where there is such an excess of nitrogen in the blood from diving deep that if any animal comes up too fast, as the pressure decreases dramatically the excess gas is no longer held in solution in the blood but forms bubbles that increase steadily in size as the animal approaches the surface. Humans are particularly sensitive to getting bent, since we have not had any evolutionary selection pressure for this gas exchange. That is why the dive computer is god. So is the back-up dive computer. Truly a case of the more gods, the better. The wrong dive profile when going deep could be fatal if one of those air bubbles gets big and is located somewhere important like the brain. But sea snakes have no such concerns. However, in terms of the huge amount of energy expended while swimming, they are more like fish than snakes. They burn through energy like no other snake and therefore must eat almost daily. They only live four or five years due to this turbo-charged metabolism. Live fast and die young.

I was up very early to watch the indigo blue of the pre-dawn sky being pierced by an array of orange flares. The tropical seas are my favorite places to do field research; I never tire of watching the myriad fish on the surface or down below. Fish of all colors dance the night away in the waters, some doing a tango, some a rumba, and some are obviously the two-left-footed fools of the fish world. Two one-and-a-half-foot-long mullets glided on the surface, their white lips against their blue-green bodies making them look like sharks all tarted up for a night out—perhaps a quick bite and then a show?

We were filming the *Yongala*, a deep wreck teeming with life. The upper deck was shimmering with neon-bright colored fish doing the flamenco for each other. Lurking in the shadows were the mottled muggers like barramundi and groupers. Going deeper, the predators were larger but fewer, until only a few behemoths were atop the delicately balanced food chain. But the one animal that no one messed with was also the stealthiest. The long and solidly built olive sea snake is six feet of muscle terminating with a broad head that contains stout fangs good for hole-punching through fish scales to deliver a rather large amount of extremely toxic venom. The venom is particularly devastating to fish, yet potent enough to hammer a human.

While olive sea snake bodies are a flat grey with a tinge of green—hence the common name—the heads themselves are a dusky orange. Their eyes are surprisingly small, indicating that in the darkness of the deep they rely on other senses for their hunting ability. I had fun with another fascinating adaptation of theirs: the ability to sense light with their tail. This strange feature allows them to sense whether their tail is sticking out from whatever piece of plate coral they have sought refuge underneath while sleeping. This is important because even though they are highly venomous, they are still vulnerable to predation from large fish such as Maori wrasse or tiger sharks.

During one dive, one of the other team members gestured for me to look down. I glanced down and, much to my surprise, saw I was standing on a five-foot-long olive sea snake that was looking at me with nothing but benevolence in its eyes. I finned upwards to release it and this snake promptly attached itself to our group. It followed us through the rest of the dive like a very devoted but scaly puppy. It would swim right up to the face mask of a diver, stick its tongue out a few times to try and determine what this strange creature in its environment was, and then placidly swim off to inspect another diver. It was truly an auspicious beginning to the trip.

The fish in these areas have been subjected to a very strong selection pressure, particularly highly specialized fish like sand eels, which are ruthlessly hunted by the asymmetrical elegant sea snake. With a long, narrow head, a very long, thin neck that accounts for half of the up to six-foot-long body, and a very muscular last half, they are fettuccine with fangs. The selection pressure has been so extreme that sand eels in areas where sea snakes occur are much more resistant to sea snake venom than sand eels where sea snakes do not occur. They still get predated on, but it would have been carnage when the sea snakes originally appeared in the oceans after the first live-bearing Australian elapid snake decided to have a "sea change." Anything that makes a hole and lives in it like a hobbit is specifically targeted by sea snakes. Out on the sand flats I was entranced, watching the elegant sea snakes glide effortlessly along. The irregular dark blotches all along their tawny bodies provided the perfect camouflage. When immobile, they disappeared against the dappled sunlit sandy bottom.

I collected snakes by putting my hand into a mesh dive bag, grabbing a snake mid-body with the mesh, then using the other hand to turn the bag inside out. This resulted in the snake being on the inside and me on the out. As sea snakes breathe air, I could not hold on to them for the entire dive, since they might drown. Even though sea snakes can hold their breath for long periods of time, I had no way of knowing how long it had been since their last breath, plus any struggling during capture would use up vital oxygen that much faster. Thus, I had to assume that the snake was going to need air shortly. However, it is unsafe diving practice to keep going up and down from the surface, called "yo-yoing." I settled on a very simple way of accomplishing what needed to be done: I would partially fill a balloon with air, place it inside the bag and send the snake up to the surface that way. As the balloon traveled up, the air would expand greatly, so I was careful not to overfill it so that it did not pop on the way up. On the surface, one of the crew members would zip over in a small boat and collect the bag, to put it in a shaded water tank back on the main boat.

Most of the dives were no deeper than 65 feet, but the last one of this particular day was to 125 feet, a depth where nitrogen narcosis can start occurring, especially if a diver is tired. Oxygen makes up a small percentage of the air we breathe, with nitrogen constituting most of the remaining 80 percent. Jacques Cousteau spoke of nitrogen narcosis lovingly as "rapture of the deep:" the intoxicating effects nitrogen produces when air is breathed under pressure at depth. The deeper one goes, the higher one gets. Nitrogen dissolves into the fatty material that covers nerve cells and subsequently interferes with the transmission of nerve impulses. On the deep dive, I felt the effects of the nitrogen: my reflexes became slower and my thinking not as clear. This was demonstrated nicely when I neglected to close a bag while I was blowing up the balloon to float it up to the surface: the snake escaped halfway there.

By this time, the weather had shifted dramatically and there was a major storm developing that was predicted to head straight in and crash into the heart of Townsville. It had already been given the name of Cyclone Tessi—which made it sound like an emotionally unstable super ex-girlfriend was out to seek revenge. It had total bunny-boiler written all over it. So we quickly made for shore, parking the boat in the Horseshoe Bay marina on Magnetic Island and grabbing the last ferry back to the mainland. The cyclone struck six hours later. We were with the director of the shoot, Russell Kelly, in his apartment, sheltered from the winds by the massive Castle Hill, and saw the eye come over Townsville. Strange moonlight bathed the buildings for a short period of time. The palm trees went from being bent ninety degrees to standing still and straight. Then, as the other side of the eye approached, the winds slammed in again and the trees bent once more, this time in the opposite direction.

Daylight brought scenes of devastation, including trees impaling houses, boats on land and cars in the ocean. The only place with power was the local pub, which quite naturally had the latest and greatest in man-toys: twin automatic-switch large diesel generators. Life continued for the pub without disturbance but with a significant monopoly on the food and drink business. Prices were normal and the food and cold beer plentiful. Construction-worker mates of the proprietor had turned the fresh hole in the roof into a glass-sealed skylight by 10 a.m. Each of the four workers walked away with a friendly handshake and a slab of beer. As torrential rain and strong winds were still lashing the city, I was wearing my blue-lens swim goggles to protect my eyes, looking to the locals not like a scientist taking child-like delight in cataloging a natural disaster but more like just another stoned, shaven-headed male tourist from Sweden. Weird but harmless. Fine, I could live with that.

Luckily, we had completed the ocean filming sequences—the water quality would be blown out for over a week after a storm like that—so we headed up to Innisfail to film snakes kept by a local snake keeper. We had been there less than two minutes when he was bitten. He had two albino death adders in a cage with a piece of cardboard as a low-tech divider. Both snakes were on the same side, one of them hidden under some newspaper. As he was competently hooking the one he could see, the other struck as the shadow of his hand passed over, connecting with both fangs and leaving twin holes in the newspaper. The cameraman hadn't even put his gear together, so everyone was entirely unprepared for this.

I rapidly pulled out my first aid kit and wrapped his arm in a pressure-immobilization bandage to slow the spread of venom. We then bundled him into a car and raced for the hospital, which was no less than twenty minutes along the steep, twisting roads. I had my ampoules and needles ready in case he went into allergic shock. While that did not manifest, a drooping of his eyelids and deepening of his voice indicated that the neurotoxins were start- ing to exert their chilling effects. An Irish female locum doctor staffed the small local hospital that day. The stereotype of Irish women being willowy and beautiful held true: her long, tangled, reddish-brown hair offset emerald eyes. She wasn't camera-shy in the least as cameraman Richard Fitzpatrick gazed up at the scene, his massive lens like the Eye of Mordor.

By now the snake keeper was displaying all the symptoms of severe neurotoxicity, particularly in the left eye, which was now immobile and pointed sharply outwards. Australian antiven- oms are known to be the best in the world, but even so, he was premedicated against shock with a shot of adrenalin to the stomach, as is protocol, with each injection feeling like yet another giant bee sting. Administration in this case turned out to be prescient since he immediately reacted violently to the antivenom and went into allergic shock in reaction to it. His skin erupted in large hives. The doctor took pity on him and offered to scratch wherever it itched the most. With the least paralyzed of his arms, he clumsily gestured towards his crotch. She looked straight into the lens and said, in an even tone, "I'm not touching that." She then left the stage.

Once the snake keeper was stabilized, the medical team made the decision that he needed to be cared for at the Cairns Base Hospital since he was having severe breathing difficulties. There was one complication, though. Another cyclone had formed and was potentially going to cross the coast near Cairns. Acting quickly, we loaded him on to the helicopter, which took off with great urgency. They landed four hours before the effects of the next cyclone were felt.

Not long after, I was off to Broome to meet up with an Animal Planet film crew for a documentary on sea snakes at Ashmore Reef. I was guest-starring on the show of a herpetologist I shall refer to as the Red Dwarf. His program was an authentic one in that he did not stage shots, unlike most on this station, with its plummeting standards. Instead, all the captures were filmed as they occurred. This strict policy of no-reshoots unfortunately meant that the programs were plagued with bad camera angles and poor light. However, unlike most of the other hosts, Red Dwarf had not appeared from a white-trash alternative universe via a wormhole.

Ashmore Reef is located so far offshore that it is much closer to West Timor than to continental Australia. Luggage check-in at an airport is typically a mundane affair; however, this time the attendant at the airport's Ansett counter looked astonished at the mountain of gear of all sizes and shapes that I presented. There were nets, snake hooks, a very large first aid box (a necessity whenever I go out in the field), a dive bag, a cool-box that was unusually heavy (it contained my dive weights as well as cold packs) and two rucksacks. All up, six pieces with a total weight of nearly 220 pounds. The staff member checking me in gave a shudder at the mention of snakes, and promptly lost interest in my excess luggage, expeditiously and adeptly checking it all in, wishing me the best and obviously hoping I would hurry along. To my eternal gratitude, no mention was made of excess baggage fees. I thanked him profusely and scampered away before he could recover from the nasty visions of snakes and reconsider the matter at hand.

Upon arrival at Broome Airport I made straight for the boat. The vessel we would be taking was the one-hundred-foot-long *Kimberley Quest*, a sumptuously fitted-out boat that promised to be very accommodating for the thirty-six hour voyage to Ashmore and Hibernia reefs. Ashmore Reef National Nature Reserve (now known as Ashmore Reef Commonwealth Marine Reserve) is located in the Timor Sea, approximately 520 miles west of Darwin and 380 miles north of Broome, or, in more practical terms, about 93 miles south of West Timor. The reserve consists of three small islands, a large reef shelf, and 225 square miles of seabed. This remote reef system is a critical stepping stone in the transportation of nutrients from the rich reefs of Asia to the reef systems located along the Western Australian coast. Ashmore enjoys the highest level of protection afforded by the Department of Environment. It is a special place, with the greatest concentration of sea snakes found anywhere on earth. In addition, many of the local species are only found there. The research undertaken during the filming would provide further data about the relative abundance and uniqueness of the sea snakes we encountered.

The water rapidly changed from navy blue to turquoise as we approached. The water visibility was around one hundred feet and the water was at a delicious eighty-three degrees. We geared up for our first dive of the trip; this dive was not for research, but rather an assessment dive, designed to provide time for the divemaster to assess each person's relative dive competence, while also allowing each diver to do last-minute gear checks. This was particularly important, as Red Dwarf had only recently been certified, with his entire dive experience being within a single quarry in the United Kingdom.

The shallow lagoon was teeming with life. Large schools of stingrays glided gently along while birds soared and dived in the skies. In addition to hosting an unprecedented number and diversity of sea snakes, Ashmore is also home to important rookeries for several species of birds. The snake density was truly on a legendary scale. We caught twenty snakes in five minutes. All belonged to three species that I had never seen before: the dusky sea snake, Dubois' sea snake, and the leaf-scaled sea snake. All three species are smaller relatives of the olive sea snake. The dusky sea snake looks essentially like a small brownish olive sea snake, although not as heavily built or laterally compressed. The Dubois' sea snake was much the same in build, but black with scales outlined in white, giving it a reticulated appearance. The third species was something radically different. The scales were pointed and heavily overlapping, looking like glued-on leaves. The leaf-scaled sea snake was truly a gorgeous animal.

During the surface interval after the dive, I went for a long snorkel to assess the area and came across a breeding pair of turtle-head sea snakes. This curious species is actually moving away from being venomous. The venom glands have shrunk, as have the fangs since these snakes feed only on fish eggs, which they scrape off the rocks using specialized scales on their chins. The males are smaller and can be almost jet black, while the females are larger, as is the case with most sea snake species, and much lighter in pattern. The male also has a thick, pointed scale on its upper jaw, which gives the effect of a turtle's beak, thus giving these snakes their common name. The male, as I was lucky enough to observe, gently pokes the female in the neck region with this specialized scale during the courtship ritual.

Our plan was to spend an entire day on the outer face of the reef looking for a much larger species of snake—the Stokes' sea snake. This species was the sole reason we had been wearing 5mm wetsuits, despite diving in the tropics. This snake has fangs long enough to puncture a thinner wetsuit and a large specimen could possibly envenomate even through wetsuits of this thickness. Truly a formidable adversary.

The first dive of the day was very deep, with the maximum depth predicted to be at least 130 feet. Thus there were two competing, but interlinked, concerns: rapid depletion of our air reserves and nitrogen narcosis. The dive started uneventfully enough, although Red Dwarf was very slow to equalize his ears, so I used up ten precious minutes of bottom time and a fair bit of air waiting for him. Once the team was assembled at the bottom, we proceeded to methodically search the area. A large olive sea snake was captured soon after we set off. However, there was no sign of any Stokes' sea snakes. Further searching of the area yielded no more specimens. As we approached a large bommie—a huge boulder-like coral structure around which marine life is often plentiful—my pre-set air-level alarm went off, indicating that I was getting low on air.

I signaled to the divemaster that I was low on air and proceeded to make my way up to fifty feet with another team member in order to do my safety decompression stop before leaving the water. A minute or two into this stop I noticed with alarm a large quantity of bubbles coming up from below. Seconds later, Red Dwarf and the divemaster came rushing past me like a snake in a mesh bag with a balloon. Red Dwarf had his arms and legs wrapped around the divemaster, preventing him from dumping air from the BCD (buoyancy control device) and thus stopping this dangerous uncontrolled ascent. There was nothing I could do but continue my safety stop and watch from below as the crew boat sped over from the *Kimberley Quest*, picked up Red Dwarf's limp body and then sped off. I watched the divemaster prudently make his way back to eighty feet to try to recompress the nitrogen gas in his blood in an attempt to avoid getting bent.

Upon surfacing, we signaled to the crew to come and pick us up. As we headed back to the *Kimberley Quest*, we pestered the boatman as to what had transpired. It was quite simple: Red Dwarf had run out of air. He had been doing pieces to camera while wearing a full-face mask linked up to an audio recorder that was synchronized to the camera. While doing this narration he had used up his air much more quickly than would normally be the case, which would also have increased the effects of nitrogen narcosis. The divemaster had to remove Red Dwarf's mask in order to let him buddy-breathe from his air supply. However, having had a near drowning incident a few years earlier, Red Dwarf was very anxious about water, so when the mask was removed and water hit his face he panicked.

Back on the boat, Red Dwarf was receiving emergency oxygen to prevent decompression sickness. Oxygen speeds the removal of excess nitrogen from the body. He was looking a bit pale but other than that was fine. The expedition doctor decided that Red Dwarf would stay on oxygen for a full hour and that the divemaster would, as a precautionary measure, also receive oxygen. Meanwhile, the assistant divemaster checked out Red Dwarf's equipment and found all to be working perfectly. Thus, equipment failure was ruled out as a cause. This left as the only possible remaining scenario Red Dwarf's getting disoriented due to nitrogen narcosis at that depth, with the added fact of his using excessive air for his talking to camera. The loss of mental sharpness meant that he had not noticed the rapid depletion of air in his tank. The incident occurred through no fault of the divemaster. These things are preventable and shouldn't happen.

Strict dive policy dictates that adequate supplies of pure oxygen for emergency use must be present prior to the commencement of any dive. However, all the emergency oxygen supplies had now been used. The expedition was therefore over. No arguments. I was quite disappointed about not achieving the goal of catching a Stokes' sea snake for the research, but remained philosophical about it. I would have to keep searching and find one another day. The documentary company had enough material to make an entertaining film, so all was well at the end of the day.

A few months later I was off to Niue to film sea kraits. Unlike sea snakes, sea kraits are egg-layers, having evolved from the egg-laying kraits in Asia. To get to Niue I had to fly first to Auckland, New Zealand, to meet up with the film crew. On arriving I was instantly struck by the realization that no matter how many layers of quick-dry clothing are worn, a person will still shiver in the cold winter. I had totally forgotten to take into account just how cold Auckland is at that time of year. After a five-hour layover we flew to the much warmer Tonga. The landing at Fua'amotu Airport was followed by a scenic-route tour of the island, taking the long way to Nuku'alofa, where we would stay the night. The tour was informative—or at least it would have been if it weren't being conducted in the middle of a moonless light. The guide even said, "Over here is the residence of our King Tāufa'āhau Tupou, we love him so. If it was light you could see the many hedges on the property." Obviously, this was another island-economy job for a cousin of someone in the government. An early rise the next day was followed by a more direct drive to the airport. As my gear was being loaded on to the security belt, I said to the female worker, "Let me get that, it's heavy." She gave me an amused look and then effortlessly bicep-curled a very heavy rucksack. Her flexed bicep was bigger than my calf muscle. Okay, righto. As you were.

As we boarded the small plane, I saw the crew forcing more bags into the already crammed hold. For small, old planes such as this one, this could lead to loss of control from overloading. As we approached Niue, we were buffeted by the very strong winds characteristic of the island. From the air, I could see that the wind had generated huge, unsurfable breakers on the windward side. But it was the unsteady twisting in the wind by the overloaded plane that really caught my attention. Somehow, we landed unscathed.

Niue is a tiny limestone atoll north-east of Tonga and south of Samoa. This speck in the ocean has true blue-water dive visibility, but over corals, not open ocean. Dive visibility of over 245 feet is not uncommon. The species of sea krait we were after was known locally as the katuali. It is not the longest of sea kraits, nor is it the shortest, but it is one of the most robust due to the thick layers of muscle used to propel it through the strong surf and currents. These same conditions made diving very challenging for us. The extremely strong surge that followed each massive wave would suck us back twenty-five to thirty-five feet and then sling us forward again, which made for many collisions with the coral and rocks.

This was particularly the case during the Bubble Cave dive. The cameraman Pete West and I entered an opening in the rocks at about sixty-five feet and swam along it for about one hundred feet before rising up into an air-filled cave under the island. During the initial swim, we were smashed repeatedly against the limestone walls, accumulating numerous bruises and one cracked lens. Once above water in the cave, we took our scuba gear off. The water level would rise and fall ten feet with each fresh surge, causing the air to condense into a heavy mist in front of us from the pressure of the rising water, and our ears to snap-and-pop equalize as each subsequent rapid drop in the water level caused the pressure to precipitously fall. The only access to this cave was through the tunnel, but it was obviously well known to katuali. Scores lined the cave floor and their banded bodies hung off the walls like the world's deadliest Christmas socks for Santa. Large females crawled high up into areas well above the level of even the highest of tides. It was there that we found them laying their eggs. We filmed them and then sat back entranced, watching this precious sight. Soon it was time to return to the outside world. The trip back was as much of a washing machine spin-cycle smash-up as the way in had been.

We took a day's break from scuba diving to lick our wounds and let our battered bodies heal up, occupying our time filming land scenery and life in the shallows. The island hosted giant coconut crabs, creatures that must have been the inspiration for the "face-huggers" in the *Alien* movie franchise—giant, slow-moving crustaceans that were as tasty as they were bizarre-looking. Katuali were abundant all around the island, both in and out of the water. Giant moray eels were also very much in evidence in the surrounding sea. One especially massive specimen had been nicknamed

Godzilla by the locals. It hung out at the pier and terrorized all who came into the water. It had never attacked anyone, but its size alone was fear-inspiring. Solid orange, it was almost ten feet long, with a head that was almost eleven inches across propelled by an even more massive body. The first time I saw it, I thought it was a lemon shark cruising the sand flats. Then I realized it was actually the biggest moray eel I had ever seen.

When Niue was hit by a cyclone the previous year, the sole pier sustained significant damage. The New Zealand navy came in to repair it and the entire repair team bailed out of the water at the first appearance by Godzilla. The locals reassured them that it was a gentle giant. Sure, it could easily kill a person, but in its infinite grace, it chose not to. The team returned to the water with trepidation and anxiety. But sure enough, Godzilla swam by harmlessly several times each day and inspected their work efforts. The workers were sufficiently motivated to complete the repairs with uncharacteristic alacrity, efficiency, and speed!

Overnight, torrential tropical rain inundated the island but it was clear and sunny the following morning. However, the huge amount of fresh water running off the island created an emulsion layer, with the less dense fresh water suspended on top of the dense salt water. The fresh-water zone was clear but blurry as it bounced and ran along the salt zone. Once we descended below the fresh and into the salt, the visibility returned to the crispness characteristic of Niue. We spent the day filming the curtains of katuali as they swam to the surface to breathe and then back down to hunt in the channels formed in the coral by the powerful waves and current. They were hunting eels in particular, and seemed to have a special affinity for the blue-colored ones. So much so that they zeroed in on all things blue—including my flippers, which just happened to be electric blue. They would follow me like lovesick puppies giving the occasional test nibble with their deadly fangs.

Over the course of the day, in my excitement at diving with these animals and with the unusually clear water not giving me a murky depth indicator, I was negligent in staying within proper dive profile standards. I was particularly guilty of yo-yoing. A proper dive profile commences with the diver first going down to the deepest depth, and then coming up to shallower levels over the course of the dive. I was swimming alongside sea kraits, lost in my happiness, and yo-yoing continually—going down to eighty feet and then back up to the surface, keeping pace with the snakes. This sharp, frequent change in pressure is particularly dangerous and can easily result in a diver getting the bends. Which, of course, I did.

The next two days were wiped out by another tropical storm, which was convenient since I was in no shape to dive. My nerves had a weird sort of electrical crackle. Things were not right in the neuroscape of my body. I crunched painkillers and hoped it wasn't too bad. I was kind of in a shit-out-of-luck situation otherwise. The film crew had not been attentive in their planning, and there was no hyperbaric chamber on the small island, which is understandable, but also no pure oxygen, which was unacceptable. Flying me back to Auckland was not an option since going up in a poorly pressurized small plane would cause the air bubbles in my blood to expand even more. So there was nothing for me to do but lie in bed stoned on painkillers, listen to music like The Prodigy's *Breathe,* and try not to die.

Once the weather cleared and it was safe for me to dive again, I had a truly magical experience on the last day of filming. We were hanging off the side of the boat, resting after another physically draining dive, battling the strong surge. I had already taken off all my scuba equipment but still had my flippers on. Looking below me, I noticed that the bottom was suddenly a lot closer. But it wasn't the bottom at all. It was a humpback whale swimming directly beneath me. What instantly struck me was not just how long it was but how massive it was. The pectoral fins also appeared to be almost as long as the body itself. I grabbed a mask from the boat and took off after it. Even though it was swimming at a very leisurely pace, I struggled to keep up. As I looked upon it with wonder and a feeling of glee, I noticed that its abdomen region was bulging. It was a pregnant female. I have never felt such an ethereal, magical feeling. All my senses were abuzz with wonder and delight. We were in fairly shallow water and as we drifted over the deeper water, I took a deep breath and did a free dive to about sixty-five feet to stay with it as it started descending. But then it gave a mighty kick with its broad tail to power itself into the deep. The turbulence in the water sent me into a reverse somersault, with my neck cracking from the abrupt motion. For a second or two I blacked out. I came back to consciousness rather disorientated. Luckily, the buoyancy imparted by my wetsuit was pulling my limp body upwards. I made a few uncoordinated kicks to speed my rise and broke through the surface with a mighty gasp of air to fill my burning lungs.

Months later I saw the final version of the film, after it had already gone to air, and had decidedly mixed emotions. While the scenery came across as stunning and the dive scenes were compelling, there were sequences that I had not been involved with that were complete fabrications. These were the scenes regarding how katuali feature in the local culture. Venomous animals typically have a profound impact in shaping a culture and mythology—for example, cobras in the Hindu religion. However, in Niue katuali were uncharacteristically not a part of the indigenous beliefs. So the director had the locals on camera completely making things up as they went along. One woman went so far off the reservation that she ended up with these weird stories about katuali that were even more fetish-laden than the Greek mythology story of Danaë, who was impregnated by the god Zeus when he came to her in the guise of a "golden shower."

The "local flavor" footage was even more obviously contrived in a sequence involving a fisherman who they filmed banging on his small outrigger canoe with the oar to appease the snake god and thus have a bountiful catch—ignoring, of course, the small detail that violently pounding on the canoe would scare away the fish, not attract them. But then, even more absurdly, the sequence continued with him putting bait down with a hand line, and then cut to him triumphantly returning to shore with a six-foot-long marlin about as long as his small canoe. In reality, it was a dead one purchased from the island fish market that very morning. This was my first real experience of the duplicitous editing of such programs. Even though I played my part straight, the assembly and edit were out of my control.

A couple of months later, I headed out with a film crew from National Geographic for a new series called *Snake Wranglers*. Our quest was to finally catch a Stokes' sea snake. This time I was searching at the tip of Cape York in Queensland, in a mining town called Weipa. From the air, Weipa was an interesting dichotomy: a combination of lush green bushland and red open sores from the mining operations. Stepping onto the tarmac and taking the first breath was an unusual experience. It was extremely humid, yet the air was saturated with a fine red dust. The result was essentially airborne red mud that instantly coated the lungs. As we waited for our luggage to be unloaded, I began to wonder what industrial hell I had arrived in, and what effect it would have on the sea snake catching. Would we catch anything at all? And if we did, what would they be like? Would a one-hundred-foot chemically mutated sea snake storm out of the water, demanding to talk to whoever was in charge about some two-headed babies it had back in its lair?

We were shown around the mining town by two thoroughly affable Aboriginal Comalco employees named Warren and Rocky, who were quite obviously completely thrilled with the idea of taking a couple of days off from their real jobs to squire us around as we looked for land snakes, when we were not out on the water catching sea snakes. Joining us also was a local snake-mad teenager named Lauren Collings. They showed us the different aspects of the mining operation, including the trucks used to move the bauxite ore. I'm not sure if calling these things trucks is the proper word, considering that each tire was about ten feet in diameter and this mechanical beast had about ten of them—the entire thing was about fifty feet long. At Lorim Point was the ship-loading pier, with all sorts of mechanical contraptions on it. At night it was lit up like a Christmas tree and was stunning to behold. All in all, a very impressive operation. Talking to my guides, I learned more about the mining operation and that it was actually surprisingly sustainable. They scrapped only the first three feet of soil and then replanted the area when done. As trees grow so fast in the lush tropical environment, areas mined as recently as fifteen years ago were almost indistinguishable from virgin terrain.

An unseasonal couple of days of monsoonal rains interfered with our sea snake catching, but did bring out a few million amphibians. While some were indigenous species of frogs, the vast majority were that scourge of the environment, the cane toad. The effect of the cane toads on the local wildlife was quite evident. No quolls were spotted during our entire trip, despite historically being a very common species in the area. Several nights of road spotting revealed not a single snake, only cane toad after cane toad. Locals confirmed that this was not an anomaly and that the quolls were very scarce, as were many previously common species of reptiles. Depressing.

With thunder booming in the distance like a maniac drummer in a heavy metal band, and lightning crackling across the sky like spider webs, we embarked from the boat ramp with a local fishing guide named Dave Donald. Dave had the leathery face of a career outdoorsman, with deep character lines earned through decades of honest dedication to a craft. The dark tan of his face was offset by light blue eyes and steel-grey hair. Like me, he was deaf in his right ear but in his case it was from the steady deep buzzing drone of the outboard motor over the many years he had been a fishing guide in the far north of Queensland. Dave had the rough, salty humor of the rural outdoorsman—another reason we became fast friends.

To illuminate the snakes in the water, we used car-battery-powered spotlights. Box jellyfish showed up as ghost-like comet shapes in the water. The abundant box jellyfish ensure that at the very time of year you would most like to go for a swim, you can't. Of course, in Weipa a swim is never recommended due to the year-round presence of large saltwater crocodiles and bull sharks. In addition to plentiful crocodiles on the banks, lounging like logs with teeth, there were numerous deadly dark shadows in the water as big sharks cruised by the boat to check us out. Each year in the early summer in the north, the box jellies come inshore to enter the mouths of rivers and creeks to spawn. Box jellyfish are the Olympic athletes of the jellyfish world, able to swim even against a current and travel up to three miles a day. They have four very primitive eyes in the multi-chambered bell. The sixty ribbon-like tentacles are arranged into four bunches, with each tentacle being over ten feet long. These invisible Furies possess one of the most devastating venoms of all. The venom from a stinging tentacle causes a pain so intense that people sometimes die just from the shock. The pain from the frying of the tissue is agonizing, feeling like long trails of acid being poured along the skin. If a person survives that, the direct effects of venom on the cardiac system can also kill. Human survivors are left with deep scarring, as if violently whipped with a thin metal rod. The clear, soccer-ball-sized bell lacks stinging cells, so I took them from the water by gently grasping the high-tech plasticine-like material and lifting them up. The tentacles were then clipped off for venom extraction later in the lab.

Each night, we spotted around eighty sea snakes in the three-hour activity period that commenced right after dusk. It was like a switch was flicked. One minute there were no snakes in sight; the next, there was one resting on the surface every few hundred feet. We were successful in netting about half of these. The method of catching was simple in theory—illuminate a snake with the spotlight, bring the boat to it and scoop it out of the water using aluminum prawn nets. Simple, that is, except that no one had informed the snakes that they were to placidly wait on the surface to be collected. Instead, they dived rather quickly after being lit up. Our greatest success was with individuals who were curled up on the surface, tied up in a nice little knot while swallowing their latest fish victim.

The first night out, we got on to the biggest elegant sea snake I have ever seen. At eight feet, it was also the longest sea snake of any kind I have ever seen. Half an hour later, a shorter but more massive object appeared in the lights ahead of the boat. It was a Stokes' sea snake, and a big one at that. It was around the six foot mark and its mottled body was thicker than my upper arm. It was diving as we approached, so I leaned way overboard, thrust my hand deep into the water and just managed to get a hold on the tip of its tail. I yanked it onboard—I finally had one! The head was bigger than my clenched fist and the fangs were less than half an inch long. The venom yield was tremendous. The 1 cc of venom contained over 150 milligrams of venom protein—far more than was needed to do a full battery of assays on it. I felt deeply satisfied, not only because it was the last piece in the puzzle for the sea snake study, but also because it was such an iconic animal. This was the pinnacle of sea snake evolution and it was a privilege to behold.

The next night out we caught a very strange snake. Instead of the smooth scales that make a sea snake so aqua-dynamic, allowing it to slip its way through the water, this was covered with extremely rough scales. It was quite unlike any of the thousands of other sea snakes I had seen previously. I asked Dave about the bottom conditions in the area, and he described it as packed full of a very unusual type of sharp rock that cut fishing lines like a razor. I theorized that the snake was evolutionarily selected to have such rough scales as a means of protection against the rocks that would slice up a sea snake into serpentine sashimi. I was extremely excited by now, since I knew with great certainty that we had discovered an entirely new species.

On the last night we caught something quite unusual again: a very large horned sea snake. Yet again, this snake was unlike anything I had seen before. Instead of feeling like a very firm water balloon, it felt like concrete. We were unloading the bins onto the dock at the end of the three-hour activity period when some young kids came down to check out the snakes. I was pointing out and naming the different kinds. I pointed at the horned sea snake and said, "We don't know if its venom is as different as its body, but I reckon it would kill you." The movement of my finger attracted the snake's attention, and even though my finger was almost two feet above the three-foot-long, thick, muscular sea snake, it had a go at biting me—and succeeded. Many land snakes would struggle to strike straight up, let alone any other sea snake I knew. But this one did it effortlessly.

I watched in slow motion both sides of the snake's head go concave as it emptied its venom into the meaty part of my left thumb. Luckily sea snakes' venom is different enough from their land snake relatives' that my allergy was not triggered. However, the venom effects came on fast and furious.

Sea snake venom is notorious for being very quick-acting, consistent with the snakes' need to rapidly immobilize fast-moving fish. If the fish can dart off, they're gone. Sea snakes are not fast enough to pursue for any great distance, so they are unable to track their prey the way a rattlesnake would follow a mouse across the desert. Despite the rapid administration of pressure-immobilization first aid, by the time we got to the small Weipa Hospital my face was grey and my lips were green. The world was getting very distant and my lower back was hurting something fierce: I was feeling the effects of severe neurotoxicity, and my muscles were being severely damaged by myotoxins, which made my urine look like Coca-Cola. By then, I was also in extreme pain. We had brought a vial of our own sea snake antivenom, as usual, and the local hospital also had a vial. Within ten minutes of arrival, the first vial was administered, and the second followed an hour later. This reversed the nerve effects and halted the muscle effects.

But the damage was already done. Back home, for a week I could barely walk and even short steps defeated me. If I put a backpack on, my back would sway under the load. For two weeks my body felt like I had competed in an ironman triathlon without training. It took me a month to be pain-free. I waited another two weeks and then resumed swim training. On my very first lap, when doing butterfly stroke, both of the rotator cuffs in my shoulders disintegrated—they were torn halfway through. For six months, I could not lift either arm above my shoulder. If I moved my shoulders in certain ways, it sounded like gravel grating against more gravel. My shoulders were now permanently wrecked and my swim competition career had come to an abrupt halt. As part of the recovery process, I spent some time on beaches. The motion used in sea kayaking did not aggravate my shoulders, so I would kayak for a few hours each morning, and then lie in a hammock chatting the day away with friends and strangers and eating lots of fresh fruit and seafood.

It was then time for me to pack up and move to Singapore to take up a research position at the National University of Singapore, as the opportunities in Australia at that stage were few and far between.

4
SINGAPORE SLING

I was not entirely happy about leaving the wildlife wonderland of Australia at the beginning of 2001 for the starkly contrasting concrete jungle of Singapore. Its lush rainforest had been chopped down and the entire island almost entirely paved over. All for the worship of money, and the prevailing mentality was that nothing would get in the way of that. The draconian approach of the government was immediately apparent with not-so-subtle signage in the airport, such as WARNING: DEATH FOR DRUG TRAFFICKERS UNDER SINGAPORE LAW.

As it is almost directly above the equator, Singapore was so humid that I felt like leaving the apartment wearing my scuba gear. The apartment was yet another anonymous cell in one of the giant beehive hells that the Housing and Development Board constructed in kit form all over the small island. Most white Western expats lived in high-price enclaves with others of the same ethnic flavor. But I wanted to save money while in Singapore, so instead chose to live in one of these massive, industrialized complexes that cost only a third of the rent of something more salubrious. The furnishings came complete with hot and cold running cockroaches—the bloody things were everywhere, but they were not nearly as cool as the ones in the cult classic film *Joe's Apartment*. No congenial Brother Ralph. Instead, these mindless cretins lurked under every plate or any flat object. I was forever nuking them in the microwave. How they got in there I have no idea, but exploded roach smells even nastier than microwaved mouse.

The air quality was absolutely shocking. I arrived right after nearby Indonesia commenced their annual torching of large swathes of primary forest to destructively clear it for the palm oil plantations. The prevailing winds carried the smoke directly to Singapore, creating terrible air pollution. The bathroom sink would have a layer of grey powder on it by 3 p.m. Socially, Singapore is a very strange and maddening place at times. It is the kind of place where the government has crack squads of scientists trying to isolate the part of the human genome responsible for bad thoughts and free will. Combined with a manic worship of money, this leads to a uniquely twisted population. It extends to all levels of society, and is immediately apparent to any traveler who has the misfortune to take a taxi in from Changi Airport. The taxi drivers rip off customers by quickly pressing on the gas pedal and letting it snap-depress. The truly infuriating part is that it is a no-win situation. Along with a low-grade case of whiplash the customer gets charged more, and the moronic taxi driver loses out too, since such rapid, repeated accelerations use up more fuel. Whenever the alien stresses of this cultural clash got too much for me, I would ring my mates Chris Hay and Tim Nias. Their strange senses of humor and empathy would decompress me greatly.

One day, I woke to the horror of the September 11, 2001 terror attacks. While not affecting me directly through death or injury to anyone I knew personally, the Twin Towers attacks were nevertheless deeply disturbing. I noted that, at best, there was ambivalence from most of the Asian population in Singapore, while certain Middle Eastern segments were definitely on the side of the terrorists. Some did not directly advocate such horrific actions, but rather pointed out the geopolitical genesis of such madness. It was reminiscent of the kamikaze pilots in World War II, also driven by religion, in that case Buddhism and Shintoism; the pilots were known as the "divine wind." But a very different ill will blew that day.

At almost the exact moment I turned on the television to find out about the terrorist attacks on the Twin Towers, I received a phone call letting me know that my mate Joe Slowinski had been killed by a krait bite in Myanmar. As the details emerged, I felt the oil and water combination of grief and anger. It was such a senseless and preventable death, brought about by the classic fuck-up cascade. A teaching case of exactly what *not* to do in the field. The perfect example of how a person could survive one mistake, but not a series of mistakes. In short, it was a total clusterfuck, where a domino reaction occurs, with a body at the end.

It was peak rain season, a difficult enough time to be in the field under any circumstances: hot, humid, and with mud that suction grips every step. This alone would cause anyone to be fatigued enough to make mistakes. However, Joe was leading a very large expedition that was plagued by poor forward planning. Inefficient delegation meant that Joe had far too much on his shoulders, leading to additional distraction. The less-than-optimal situation was exacerbated by local corruption—the items the expedition had paid for in advance to be provided in-country were either insufficient quantity or absent entirely. They were also in an area outside the permits-approved zone. The straw that broke the camel's back was that the night before the fateful event, Joe was drinking copious amounts of alcohol until the small hours. At 8 a.m. the following day he certainly wouldn't have been within the legal limit for driving, let alone in shape for working with venomous snakes.

It was at this time that a local snake catcher came and said he had a wolf snake in a bag. Joe reached in without looking first to verify the identity of the snake. Under no circumstance, even if it is my handwriting on the bag, would I reach in without confirming by visual inspection that the bag did indeed contain a non-venomous species. It is suicidal to not check, and to accept someone else's opinion that a snake is harmless, especially when the stated snake looks virtually identical to another local snake which happens to be lethal. Joe pulled his hand out with the snake clamped on to his finger. It was not a wolf snake. It was a krait.

The remoteness of the area, the lack of a doctor as part of the field team, the absence of antivenom and proper artificial respiration equipment on-site, the alcohol in Joe's system reacting synergistically with the venom, the inability to communicate effectively with a medical center, no extraction plan, being outside the approved area causing difficulty in obtaining help from the military for extraction, poor weather hampering helicopter transport once arranged, and other variables all combined into the perfect storm. The basic equation was: no antivenom + no doctor + no transport = no chance. In addition, pressure-immobilization bandaging was not immediately applied and the venom was absorbed at full speed and with maximum efficiency.

Joe progressed through much the same stages of neurotoxic effects as I had with my death adder envenomation, as the toxins paralyzed the voluntary muscles. Loss of physical coordination was the first obvious symptom, as his arm and leg muscles were steadily impacted. This was accompanied by slurring of speech, ptosis, giving a cockeyed look, and, most ominously, difficulty breathing. As they were woefully ill-equipped for such an event, the expedition members took turns giving Joe mouth-to-mouth respiration once he was no longer able to breathe on his own. Joe's sense of humor was present to the end, with him mischievously signaling that he only wanted the female team members giving him artificial respiration.

While all this was going on, a logistical nightmare emerged when help was sought from the Myanmar government. Obtaining permission for airlifting and getting hold of the antivenom was extremely complicated and time-consuming. While he could talk, Joe had said he would have refused antivenom if it had been present, since he had had an allergic reaction to antivenom previously. This is a maddening mentality all too present in herpetologists. The only effective treatment for envenomation is antivenom. Any allergic reaction can be managed through the use of injectable adrenalin and antihistamines. In any case, no antivenom was available, nor was there any artificial respiration equipment. Mouth-to-mouth respiration is sufficient for short-term, acute use. However, it is ineffective over the long term in providing enough oxygen while inflating the lungs enough to prevent suffocating fluid build-up. The only cold comfort I could take was that as the neurotoxin-induced paralytic effects progressed towards the lethal climax and he entered the locked-in phase, he would have been experiencing the same euphoria I had from my death adder envenomation. So he at least died pleasantly, rather than screaming in pain and agony.

Joe's death hit me especially hard since it was very much of case of "but for . . . there go I."

Things were getting a little too much for me, so I was very happy to retreat to my snakes. I had established a very large and diverse collection in one of the off-exhibit quarantine buildings at Singapore

Zoo. The head vet, Paolo Martelli, was a herpetologist and he had fitted out the room with a custom array of cages on all walls. The cages were still empty when I arrived in Singapore, so with Paolo's permission I filled them with snakes from all over the world for my research. We had it all, including iconic snakes like Gaboon vipers, king cobras, green mambas, Fea's vipers, and boomslangs. Most important from a research perspective was the tremendous assortment of strange snakes which had the more primitive venom delivery systems consisting of enlarged teeth in the rear of the mouth. Their venom glands did not deliver the venom as a high-pressure stream, as a cobra's would. Some of these snakes were harmless to humans, but still of significant evolutionary interest. Others, however, like the olive sand snake from Africa, were as toxic to rodents as a cobra, and thus posed an obvious human danger. Even more so since there weren't any antivenoms that would cover most of them. We had the boomslang antivenom before we got those snakes, but it took us a year to get the keelback snake antivenom from Japan. That was a far from ideal situation. By the time we obtained it, we had already milked those snakes for sufficient venom and moved them on to make room for new species.

One of the most delightful parts of this new research was starting to work with slow lorises. These adorable little creatures are the only venomous primates in the world, having a bite that is extremely painful. They also coat themselves with a secretion produced by glands on the insides of their arms that causes allergy-like effects in potential predators. The relationship between the arm secretions and the bite pain was unclear. Dissecting preserved specimens, I noted that the submandibular salivary glands were unusually enlarged, which suggested the source of the bite pain, with the arm secretion thus being unrelated. It was exactly the kind of riddle I liked unraveling.

Helping me with the research was my student Ryan Ramjan, and also Tim Jackson, who had just finished high school in Sydney and was taking a year off to live in Singapore with his parents, who had recently moved there. We formed a very tight team and had a lot of fun. We spent many hours exploring the zoo, as we had access to all areas, on and off exhibit. We would often be found at the reptile house chatting with the awesome curators, Bernard and Francis. Bernard had been bitten not long before by a mono-cled cobra, a bite that basically melted one of his fingers, leaving a scar that we examined with great interest. Knowing background details like that only makes one that much more respectful when milking the same snake for its venom. We were also regular visitors at the vet hospital, hanging out with Paolo, who was never boring because vets always have the best screw-up stories. One incident in particular stands out. Paolo had several very large pythons in residence at the vet hospital. Whenever one of the carnivore depart-ments had a spare live rabbit, it would be sent over to the hospital for Paolo to feed to his snakes. He came in one day, saw a rabbit, absent-mindedly snapped its neck, fed the pythons, and thought no more of it. Four hours later, one of the administration assis-tants came over to chirpily inquire how things were going with the check-up of the pet bunny belonging to the niece of the Deputy Zoo Director. Oops.

My usual morning routine consisted of eating breakfast of roti *prata* at the National University Hospital cafeteria with Tim, and then taking a taxi to the zoo. One morning we were running a bit late, but I desperately needed some coffee as I had been up late watching European Champions League football. Coffee in Singapore is served at a temperature that is only slightly less than that of the surface of the sun. In the equatorial heat, it takes forever to cool to a level that won't scald. As we didn't have time this day, in my sleep-deprived state I settled on a cunning plan. I would cool it down with some of the liquid nitrogen I had with me. I poured about 50 cc into the coffee, which promptly erupted into a spectacular three-foot-high brown geyser. The room full of people, several hundred of them, went instantly silent. As we rapidly packed up and headed for the exit, Tim quite rightly gave me a look that said in no uncertain terms that I had just made the biggest dumb-ass move he had ever had the privilege to witness.

Getting in shipments of snakes was like receiving Christmas presents, and was always cause for excitement, with us quivering in anticipation. One lot in particular stands out, but not for the best of reasons. We had a shipment come in from Malaysia that contained a large number of banded kraits. Kraits have a Dr. Jekyll/Mr. Hyde split personality: they are typically rather calm during the day, but absolutely psychotic at night. This fits with their behavior in the wild, where they hunt at night and sleep during the day. By the time we got back from the inevitable complications of bringing boxes full of venomous snakes through any airport, let alone somewhere as draconically bureaucratic as Singapore, it was 10 p.m.

Before we unpacked the shipment, I ran through the basic safety protocols with Ryan and Tim. First, I said, always assume that there is a snake loose in the box. Saying this, I levered a crowbar under one edge and applied force to lift the lid off. It was tightly nailed on and the wood was very thin. So instead of completely coming off, only a long pizza slice section tore off. This was enough for us to get a glimpse of a narrow black and white head looking up and giving a death stare before diving back into the dark confines of the box. There was silence as we contemplated this turn of events. This meant that there was at least one of these highly venomous snakes loose in there. How many more would be loose? We delicately removed more of the lid, snapping off successive pieces until we had enough of an opening to remove the bags one by one with tongs held by heavily gloved hands. It turned out that six more kraits were loose. The bags were decidedly not to the standard of my specific instructions. Even though they were double-bagged, the snakes had pushed through the corners, which were held together by poor-quality thread. This at least provided an opportunity to reiterate that even when snakes are double-bagged, always assume they have got out of the first bag. Just after I imparted this lesson, I opened up one of the bags still containing a snake, and resting on top of the inner bag was a krait. This was a completely unacceptable state of affairs and I vowed to never order from this vendor again.

As part of the process of obtaining venom samples for the research, several times I milked the giant king cobras that were kept in room-sized enclosures at the reptile house. While I had kept one when living in Portland, nothing had prepared me for the experience of walking into a giant cage where four very large specimens lived. The saving grace was that these snakes were so intelligent that they responded to certain behavioral cues. They establish territory by rearing up as high as they can, and they determine dominance by tapping on the top of the other's head. The one tapped would drop to the ground and slither away. I used this to my advantage by tapping on their heads to buy me time to move into position. Upon my entry, the eight foot female would always zip agilely up the rock wall to the top, scaling the ten-foot structure effortlessly and with great alacrity. The much larger males, however, would stand their ground. All the males were at least twelve feet and the biggest was over thirteen feet.

Milking these king cobras was the most physically challenging event of my career. While taipans were faster and more agile, the size and strength of the king cobras was like nothing I had ever encountered. To milk them I would wrangle them into a black sack which had an opening at the end; this was wrapped around a wide, clear acrylic tube. I would let the snakes slide along the inside until the head was just out of the tube. I would then hold a large plastic container in front of them, which was readily bitten by these massive snakes. The venom yield was staggering—enough to fill a shot glass and containing hundreds of milligrams of toxic protein.

One day I had a mangrove catsnake that was over six feet long swallow my entire thumb and then set about working its stubby rear fangs into my flesh. It took nearly two minutes to remove it. By then I was well and truly envenomated. We had already discovered in the lab that the venom was over a hundred times more potent on birds, their preferred prey, than on mammals, which they only rarely ate in the wild. So I was not too concerned about my welfare. However, in addition to some loss of balance due to the neurotoxins, I developed the most splitting headache. So I popped a few pills of the only painkiller I had handy: codeine. It promptly cross-reacted with the neurotoxins to produce the most intensely delicious high, far surpassing the death adder venom effects, but without any of the paralysis. Singapore is well known for being very harsh in its approach to drugs, so at that point I was probably the only legally high person on the entire island.

My research was concentrated on tracing the evolutionary history of a particular type of snake neurotoxin called three-finger peptides. These were the signature of venoms from cobra-style snakes, including Australian snakes such as the death adder, but were not known at that time to be in the venoms of any other snakes. Indeed, it was not settled at that point how many times venom had evolved in snakes. The prevailing theory was that vipers and elapids (including cobras) evolved their venoms separately, and that the other snakes were all non-venomous. However, this did not account for the existence of lethal species such as boomslangs in between. They were considered as special exceptions, which struck me as evolutionarily nonsensical. The line of investigation I was pursuing was that *all* of these other snakes lacking front fangs were, in fact, venomous. It was because the majority were not dangerous to humans that they were considered not to have venom—an anthropocentric bias that obscured the evolutionary reality. In fact, they were using venom to stun non-dangerous prey like frogs or geckos and the venom was delivered to these thin-skinned prey through a low-pressure system involving repeated chewing. The enlarged, grooved teeth in the rear of the mouth are sufficient to create wounds in the prey's thin skin, allowing the influx of venom.

My first big breakthrough came with a thin-bodied, fast-moving striped Asian snake called the copperhead racer (sometimes known as the radiated ratsnake). Tim, Ryan, and I laboriously milked over a hundred of these by knocking them out with an anaesthetic called Zoletil, then massively stimulating the venom glands with a chemical called pilocarpine and collecting the secretion as it gushed out. Isolation and sequencing of the dominant component revealed it to be of the same class as the typical cobra-style neuroactive three-finger toxin. Bioactivity testing demonstrated a high level of neurotoxic activity, comparable to that of death adders. This showed that not only was venom an early-evolving characteristic in snakes, but that these unstudied snakes were a rich resource for biodiscovery. This was exactly the kind of key breakthrough I needed to make the difficulty of living in Singapore worthwhile. Relief flooded my body, as I knew my professional venom research career was well on track.

The research breakthrough happened right at the point I could not handle the cultural pressure-cooker of Singapore anymore. I had accumulated enough data to write several key papers. I had also saved up enough money to live in Australia for at least a year without having to work. So I quit with no plan, other than to move back to Australia and lose myself in the outback. Happiness was the Changi departure lounge.

5
LONDON CALLING

First I was off to Cambridge to do some research at the European Bioinformatics Institute (EBI). As we were landing on February 12, 2003, I viewed Heathrow Airport from the air and noted that tanks were ringed around it. It transpired that while I was on the connecting flight from Singapore, London had become clenched in the iron grip of fear. Intelligence reports indicated that al-Qaeda agents had smuggled surface-to-air missiles into Britain, because of the looming invasion of Iraq. Once I cleared the security check—now at increased levels, with my scientific gear attracting extreme scrutiny—I met with our driver and headed to the hotel in the lovely area of Kensington where we would stay for the next couple of days. The next day at London Gatwick Airport, British police arrested a man carrying a hand grenade, and two men were simultaneously arrested at Heathrow Airport for similar offences under the Terrorism Act 2000.

Before heading up to Cambridge, I had some media to do as the European face of the National Geographic series *Snake Wranglers*.

First stop was an interview for BBC radio. My fellow guests were fascinatingly diverse. First was Philippe Petit, who infamously once tightrope-walked between the now-destroyed Twin Towers and had written a book about it—*To Reach the Clouds: My High Wire Walk between the Twin Towers*. His description of the meticulous planning appealed to the geek in me. I was hanging off the edge of my seat as he recounted the verbal battles with the police afterwards, where they were going to just haul him off to jail. But he had kept screaming that he had to loosen the wires or else there was the risk of them snapping, which would unleash a catastrophic whip action, able to easily kill or cause incredible damage as all the kinetic energy was released.

Second up was Mark "Moe" Popernack, the representative author of a book by a group of Quecreek coalminers from the US state of Pennsylvania. Their book, *Our Story: 77 Hours Underground*, recounted the harrowing ordeal endured by the nine miners trapped in the pitch-black with icy cold waters rising around them. What struck me was the calmness with which they approached this all-but-certain death and, most amazingly, maintained their sense of humor throughout it all. I do believe that this camaraderie is what kept them alive. Through the "group fear" bonding, they did not panic, which would have used up precious oxygen. Their rescue was as much to do with the organization and teamwork of those trapped below as it was to do with the superb efforts of those above. Moe shared how his wife said she would leave him if he went back to working in the mines; it was something that gave me pause for thought about my very dangerous career and the emotional toll it takes upon those close to me. Both Philippe and Moe gave me inscribed copies of their books, with the former writing: "To Bryan, an explorer of life's marvels. Keep searching!" while the latter wrote: "To Bryan, it's been a pleasure to meet a man who has such love and dedication to what he does. May God bless you always." These books are on proud display in my library.

Last, but certainly not least, was Mariza—an extremely talented singer who was making it her mission to resurrect a Portuguese folk music called fad. She sang so very hauntingly, in a strangely soothing manner.

Then it was off to the *Richard & Judy* show, along with a feisty Egyptian cobra I was going to milk live on-air. We had to film it in a separate room deep in the bowels of the cavernous Cactus TV studios. This went off without a hitch and then I had a very pleasant on-air interview with this hilariously irreverent couple. But the instant the cameras stopped rolling, Judy vaporized two glasses of wine in record time. I guess the cobra had made her a bit anxious!

Monkey-dancing over, I was on my way to Cambridge to stay at the lovely old stone house of my Finnish friend Heikki Lehväslaiho and his delightful family. I spent the next day at the EBI loading files onto the very fast computer cluster. It took less than seventy-two hours to spit out data that would have taken an ordinary desktop computer months to grind through.

While this was underway, I received a phone call from Australia. Chris Hay had been badly bitten by a large mulga snake while in the middle of the Barkly Tableland. He'd been bagging it up as part of our ongoing research, covered by my scientific permit, when he made the mistake of passing his hand between the bag and the headlights. The snake saw the shadow and struck, getting Chris solidly across the hand and giving it a sustained chew, driving the venom deep. Chris's field partner drove at high speed to the Barkly Tableland Hotel, where the bar staff stretched the now very affected Chris across the pool table. He promptly vomited all over the table. It was certainly not the first time this pool table in a very rough outback bar had been puked on—just the most unusual set of circumstances. Traffic was blocked on the Barkly Highway so that a Royal Flying Doctor plane could land. Chris was quickly loaded onto it and flown to Mt. Isa Base Hospital for further treatment. I had a feeling of helplessness and déjà vu. Was this going to be Joe all over again? Was I going to lose another mate to snakebite? Throughout the dark and lonely night I kept ringing Australia to check on progress. Twelve hours post-bite he was out of the woods. For a while it seemed like he might lose his index finger, but eventually that came good as well.

With my friend now safe and the research data mission accomplished, I started a trip across Europe for two months, with the aim of milking viper collections in zoos and those of private keepers. Some of these, the Germans in particular, had some breathtakingly rare species. The first stop was going to be Italy, but I didn't even make it out of Heathrow before the first "incident" happened. As I would be away from a source of liquid nitrogen for three weeks, I wanted to get the maximum hold-time out of the vapor shipper I was using to cryogenically store the accumulated venom samples. Vapor shippers are specifically designed to comply with the regulations of the International Air Traffic Association (IATA). High-tech foam absorbs the liquid nitrogen, so that it is no longer a liquid that can flow, but is instead in the form of a vapor. The vapor shipper can therefore be legally loaded onto the cargo section of an airplane without posing a danger. However, overfilling it so that there is liquid present in the core makes it no longer compliant. Pushing the envelope, I had overfilled it and hoped for the best, as this would give me another week of storage time.

This plan was not destined to be successful. My first indication of this was when nine heavily armed airport police came on to the plane and sternly ordered me outside, where I observed a half dozen biohazard-suited spacemen standing around the dry shipper, which was on its side. This meant that any extra liquid was long gone and it was now compliant. Doing my best to not get shot by the security services, who were understandably on alert level "paranoid," I explained what it was and that some of the vapor must have gently drifted out when the ground staff paid no attention to the clearly marked instructions and large black arrows indicating which way was up and how it should be stored. Desperately keeping a straight face, I took out the cork and turned the container upside down; I showed that it was completely compliant and that all was good. This alleviated the situation. Walking back on to the plane I gave a friendly smile in response to the curious looks from the passengers. A well-dressed white male wearing glasses seemed to trigger very few pre-set panic buttons within them.

After landing in Italy, I set about exploring Rome by first smoking a delicious Cuban cigar while at Trevi Fountain and the Spanish Steps. Due to the heightened security alert, the US consulate in Rome was under extreme protection, with orange barriers restricting pedestrian and car access, lending a surreal air as this modern incursion clashed with the ancient history. After a few days in Rome, I headed up to Florence with my mate, Mickey Bhoite. He is the delightful sort of weird character that results when an Indian is raised in Italy: spiky haircut, dark skin, and the inability to talk without moving his hands around like an orchestra conductor on crystal meth.

Florence is home to one of the oldest natural history museums in Europe, one with more reptile holotypes (the first of a particular kind ever discovered) than anywhere else in the world. Consequently, the herpetology section is absolutely massive. It was truly pickled nirvana. From there we headed down to Perugia to obtain venom from the local zoo. The keeper insisted on doing the milking himself and it was immediately apparent that he was not accustomed to doing such close contact work. Within two minutes of commencing work, the very first snake he tried to milk ended up biting him. I have never seen an Ottoman viper before, let alone studied the venom, so we had basically no idea what was going to happen. Luckily it turned out to be little more than extreme local swelling and pain, with his hand ultimately blowing up to look like an inflated surgical glove.

With this part of the Italian journey coming to a premature end, we decamped to Tuscany for a few days to enjoy the fine food and wine before piling into a campervan for a wild road trip across Austria, including several snowball fights along the way. We then traveled to Germany to arrive for the reptile expo Terraristika in Hamm the night before it was to start. Appropriately enough, the Rammstein song "Feuer Frei!" was playing as we pulled up. This was the first time I had attended this famous event. Reptile aficionados had gathered from all corners of the globe to sell their wares. There was a special, well-controlled room specifically for venomous species. Searching the room carefully, I found a number of species essential for my research and purchased them.

I also enjoyed many chats with other reptile lovers from a wide variety of countries. It was particularly interesting to talk with them about how the rise of the Internet had changed the face of the scene. The earth was now flat and close rather than being hidden in the curved distance. The Internet had definitely caused an explosion in reptile keeping. It also facilitated the sharing of experience and knowledge.

Living up to their meticulous stereotype, it was invariably the Germans who quietly cracked the riddles of hard-to-keep species. Zoos replicated their experiences, but then put out press releases about "their" accomplishments. Such actions only increased the alienation of private keepers, who became more insular and less willing to share with these sorts of professional institutions. This private/professional divide did no one any favors, least of all the animals themselves. Private keepers spent an incredible percentage of their personal income and were rabidly obsessive about the animals, conforming to the stereotype of the Japanese Otaku, but on a global level. It was almost always the private keepers who were able to not only keep alive the unkeepable, but also to get them to breed: the surest sign of success.

We also had many long conversations about the culture war being waged by animal extremists who felt no animals should be kept in captivity, least of all exotic species, but even the family dog was in their sights. Those with anti-pet agendas employed an obvious divide-and-conquer strategy. Reptile keepers were their own worst enemies in this regard, not only because some of them were public relations disasters in keeping dangerous species in unsuitable enclosures or providing criminally substandard care, but also because of the inherent splintering within the community. This only made them that much more vulnerable to draconian legislation.

At all levels, the legislation surrounding reptiles is among the most poorly drawn, or even counterproductive. This ill-considered type of legislation was the weapon of choice for such nefarious groups as Voiceless: The Animal Protection Institute, the Humane Society, and others. While they operate under the guise of concern for public safety, their motivation is the outlawing of all pets, pure and simple. The incidence of exotic-animal-related injuries is dwarfed by those caused by "companion" animals such as dogs and cats, but no legislation is proposed to ban them, other than equally misguided breed-specific legislation. If examined rationally and upon evidence, this is entirely inconsistent.

The fear of exotic animal escapes is also illogical. Feral cats cause more damage to the environment than a reptile ever could, yet cats are not being banned. Neither are hamsters, mice, and other small mammals that are also very difficult to catch when they escape. Ditto for birds, which also carry a higher risk of causing salmonella, yet are not banned. Even ordinary raw chicken meat is a greater public health concern.

Saying that animal shelters are overflowing is another smoke-screen, since very few of these animals are exotic pets. Most animals in these shelters are abandoned dogs. Yet no attempt is being made to ban dogs on these grounds. In regard to zoonosis (diseases spread by animals to humans), reptile-related cases are yet again in the minority. Nevertheless, there is an inconsistent application of the facts. No attempt, for example, is being made to ban pet birds because of disease risk. Only a very small percentage of constricting snakes reach appreciable size and thus could be considered a potential threat (though still less dangerous than a pet horse); however, included in the legislation are snakes smaller than the native snakes already crawling through yards. These snakes pose absolutely no danger to anyone.

It is evident that personal biases are coming into play; that is, it is a culture war being waged purely to support an ideology, not to combat an actual, clearly present danger. This agenda, ironically, is sure to cause long-term harm to wild animals themselves. If people only see animals in pictures, or not at all, they will not appreciate, value, or want to conserve them. Furthermore, removing from young children the opportunity to keep these magnificent creatures will kill any interest they might have in this area, and thus deny the world future generations of scientists. If I had not had the opportunity to keep a diversity of reptiles as a child, I might not have become the scientist that I am today.

After the congenial madness of Hamm, I drove to Bonn with my friend Guido Westhoff to continue milking snakes. First we stopped by his lab at Bonn University to check out his snake set-up. In addition to some of the most stunning eyelash vipers, the collection included a large number of spitting cobras. Guido was comparing the spitting distance and patterns of African and Asian spitting cobras, as these two cobra lineages had evolved their ability to spit independently. He revealed that the venom of the African species comes out in a tight spiral, able to travel up to ten feet with amazing accuracy, aimed at the enemy's eyes. This is useful on the savannah plains where predators would be visible approaching from great distances. In contrast, the Asian species lives in closed forests and therefore would only view their predators up close while hiding in a region between shadow and light. This put a selection pressure on the Asian species to evolve a strategy in which the venom came out in a diffuse spray. Basically, the difference between them was the difference between using a rifle and a shotgun.

I quickly noticed that, like myself, Guido's wife, Katja, who was his co-investigator in this study, would sneeze violently whenever in the room with the spitting cobras. They had attributed this to her being sensitive to the dust from the wood shavings that lined the floor of the cages. However, I pointed out that it was in fact far more likely that, since she was most often working with the cobras, all the sprayed venom that was drying and then wafting through the air had triggered an allergic response. This meant that, like me, she would almost certainly go into allergic shock if bitten, and it could quickly become lethal if untreated. Understandably, this caused great consternation among the research group. As they had almost finished the study, it was agreed that she would remove herself from day-to-day husbandry activities and not work hands-on with the snakes anymore. They only needed a few more weeks before the study would be complete and they could disband the cobra collection.

After this, I successfully milked a large number of vipers in collections across Germany, with a few near-heart-attack-inducing incidents. One of the scariest was in a poorly lit, clutter-filled basement, with a ceiling far too low to accommodate my six-foot-three-inch frame without me having to hunch over like Quasimodo. I was pinning a saw-scaled viper to the floor when it got away from me and sidewinded across the high-friction concrete surface and between my legs. I quickly turned to get it, in the process smacking my head so hard against the low ceiling that darkness momentarily clouded my vision. My last view of the snake was it disappearing under a box.

After several frantic minutes of searching, we found this small but particularly dangerous snake. A bite from this type of snake is notoriously difficult to treat, with antivenoms against the venom of one species, or even one locality within a broad-ranging species, being inefficient against others. This is compounded by the unknown geographical origin of many of the snakes in captivity, some of which were in fact crosses of the same species from different localities, or even hybrids between species, thus creating an immunological nightmare. Not that I would want to be bitten by any snake again, but these snakes are among the last whose venom I would ever want to have coursing through my veins.

After this debacle, I switched methods and put all the vipers first into a styrofoam-lined box. This gave me a cushioned, high-friction pinning surface that the snakes could not get out of. Ironically, they were generally much calmer in there. They would zip across the box until they hit a corner and then sit there impotently giving me a threatening look.

Another bite *did* follow not long after this. Again, not to me, which was just the way I liked it. It was to yet another professional keeper who did not want me to do the milking, but insisted on doing it himself despite having a conspicuous lack of relevant experience. The snake involved this time was a short tube of muscle called the hime habu, a pit viper species found in the Ryukyu Islands of Japan. While I had never seen one before, I could tell straightaway it would be very difficult to handle. The muscular body made its neck less defined, and the short but thick body afforded it great leverage. I pointed this out but was dismissed by the owner as not knowing what I was talking about.

Seconds later, the situation unfolded like a poisonous flower, exactly as I predicted it would. The keeper was holding the snake as I readied the milking container, despite me telling him not to remove it from the cage until I was ready. The snake pulled back, driving one long fang deep into his thumb. He dropped the snake, which promptly scooted under the nearest available cage. I looked at him and saw that he was lost in his own world, staring horrified at the large drop of blood running down his thumb. I hooked the escaped snake out from under the cage and got it back into its own enclosure. The keeper was still not on this plane of existence. As this was a species I knew I would not be seeing again anytime soon, I quickly flipped out its cage mate, pinned it, necked it, milked it and had it back in and locked away in less than sixty seconds. I then said, "Okay, let's get you to the hospital now." He came out well, with only a local flesh wound and a lot of discomfort.

It was then time to move on from Germany and into the surreal world of Luxembourg, a tiny country located within Europe but somehow legislatively special, so it does its own thing, particularly in the murky world of international banking. The first host not only had an impressive collection of rare snakes, but his wife also worked for a bank where her sole task was to give a daily churn to an account containing fifty million euro. An account that belonged to none other than Osama bin Laden. While she, as branch manager, knew who ultimately controlled the money, it was shrouded within a complex network of shell companies. Her job was to keep the money moving. If it was always somewhere else, then it was nowhere at all. I was not impressed with such a mercenary approach but kept my counsel.

The rest of Luxembourg was just as weird. For example, the pair of brothers living in a mansion with their mother, who was either an Olympic athlete in the discipline of denial or suffered from weapons-grade dementia. How else could she not know that not only did they have a large pair of alligators in the basement, but that they also housed a massive collection of venomous snakes in the upper levels. A collection that included a trio of huge black mambas, which they did suicidal things with, like taking one into the massive walk-in, four-nozzle shower and letting it loose so that they could mess with it as it slid along on the low-friction tiled surface. They would also, in a never-satisfied quest for a new high, milk it, dry the venom on a radiator and snort it. Since they mixed other, more psychoactive, chemicals in with the dried venom, any high was due to the other substances. All they were ensuring was that they, like me, developed a deadly allergy to snake venom and thus, if bitten, would die. This would basically have the effect of pouring a bit of bleach into the gene pool!

To complete the European viper venom adventure, I headed up to Norway to continue my decades-long search for the elusive arctic viper. Meeting up with me was Eivind Undheim, a hyper Norwegian who reminded me a lot of my younger self. Same shaved head, blue eyes, high, elf-like Nordic cheekbones, with the same love of venomous animals, metal music, and any other adrenaline-inducing activity—obviously the perfect partner for hunting these elusive snakes. At his suggestion, we tried a different tack, and instead of hunting in pristine forests, we targeted a farm that had a huge pile of logs and branches which had not been moved for years. This created an artificial refuge for the rodent prey of the snakes. We arrived right as dawn was spilling its light over the Oslo fjord in a very picturesque manner. We patiently took up position around the pile and settled down, unmoving other than sweeping our predatory gaze over the corpses of the trees. Before long we had success—a specimen that was almost completely melanistic, with the characteristic black zigzag pattern subtly offset by a deep charcoal grey body. Amazingly, after my twenty-five-year quest, we obtained not one but four of these rare snakes that morning. As the sun rose, each successive snake was lighter-patterned than the one before, with the last being bronze-bodied like a venomous sun-god.

We ascertained that the timing of their appearance was consistent with their ability to absorb heat, with the darkest snakes being the first to warm up and become mobile, but then also the first to overheat and seek the cool of the shade. Conversely, the lightest snakes would be the last to become warm enough to move, but then were able to stay out in the sunlight the longest in search of prey. So each part of the color spectrum was able to take advantage of different parts of the day, thus partitioning the habitat into different climate zones. This resulted in a high number of snakes, without direct competition between the different gradient types. It was deeply satisfying to finally fulfill this quest while also discovering more about the natural history of the very first venomous snake that I had ever seen in the wild as a child.

For a bit of non-snake-related time, I headed over to France and enjoyed many bottles of Chimay Grande Réserve Bleu with my friend and collaborator Nicolas Vidal. We would then decamp to our favorite Moroccan restaurant. We could also be found spending afternoons having happy chats with our mutual mate Karim Daoues at his amazing Parisian pet store, La Ferme Tropicale. I would also take long, happy walks by the river Seine, checking out the *bouquinistes* in their green-sided metal sheds and the amazing rare books and stunning paintings for sale. Rue Saint-Louis en l'Île was a frequent haven, not only for the plethora of fossil-laden natural history shops, but also the most amazing cheese shop. I would walk in, close my eyes, hold my breath while counting to sixty, and then take a slow breath in through the nose and have an olfactory orgasm. From there, I would casually stroll down the Seine to the Shakespeare and Company bookshop, my favorite book repository in the world. Walking in, the smell of old books rivals that of aged cheeses in being appealing to my snakebite-damaged sense of smell.

6
GREAT SOUTHERN LAND

Back in Australia, I celebrated with a few Coopers Vintage, Australia's finest beer. In addition to saving up enough money to cover the next year's expenses, I also had sufficient funds to buy a used Mazda 626 wagon in very good shape. On the front I rigged up six powerful spotlights in three rows of two, each pair at a particular angle to illuminate a certain perspective. Small geckos were in danger of disappearing in little tornados of smoke, and light seemed to be streaming from the ears of the kangaroos. The snake-spotting effectiveness was stellar, as the glossy snake scales reflected sharp waves of light back.

Flipping on Audioslave's "I Am the Highway," I hit the road for Melbourne, taking the coastal route at first to enjoy some nice beaches along the way and decompress after the stressful years in Singapore. Heading inland to the pleasant but boring Canberra presented the opportunity to see the beautiful radio array at Parkes that was the subject of the delightful Australian movie *The Dish*. Pushing on, I stayed the night at Bonnie Doon, the subject of another hilarious Australian move, *The Castle*. From there it was an easy drive to my new house in the Dandenong Ranges. I spent the next six weeks writing an application for an Australian Research Council postdoctoral fellowship. Once this was submitted in final form, I hit the road with Chris Hay to hunt for venomous snakes and to film a documentary.

We cut across Victoria, undertaking the incredibly boring drive to Adelaide, and then powered north from Adelaide at night across the gibber desert to stay in Coober Pedy for two nights. Hundreds of kangaroos were narrowly evaded. Unfortunately, some were not. Upon approach, groups of twenty or more would scatter in all directions, with one or two inevitably hopping straight at the car. If there was a joey in the pouch, upon impact with the vehicle it would become a detached meteoroid. We averaged one strike per night and we weren't the only ones. We saw a road train clip a cow just behind the jaw and the head spin away like a David Beckham spot kick. The headless body geysered blood a distance of at least six feet and then did a slow-motion collapse like a building being leveled. The nerve-twitch muscular spasms persisted for over a minute, with the legs moving in small random directions. Another time, a massive road train coming from the opposite direction hit three sheep while going over ninety miles per hour. The sheep exploded like white furry water balloons and the insides became the outsides.

Just after leaving Coober Pedy to head for Alice Springs, we collected a bronze-back legless lizard, a very rare species thought to exist in only a very restricted range, well away from our location. It was also thought to need the most pristine of environments, but we caught it on the highway, with mining tailings littering the landscape in all directions like piles of dinosaur dung. It was contentedly munching on a small cockroach when we came upon it.

We collected it to deposit later with Dr. Mark Hutchinson at the South Australian Museum, who was very happy and appreciative of our find. I must admit that I did not appreciate the full significance of this tiny animal, which looked like a smallish worm. Indeed, some parasitic worms I had crawl out of my butt while in the jungle were bigger than this thing.

Arriving in Alice Springs felt like entering a very special insane asylum, run by the inmates. This was seen in the oil and water interactions between the local indigenous population with whoever had just blown into town—for example, Scandinavian backpackers on a summer holiday who became all spiritual as they got stoned and made some dot painting art for the mall tourist-trap souvenir shops. They worked on machine-made didgeridoos, distinctive by their perfectly smooth insides as opposed to the random roughness of the authentic ones.

At night we would work the ranges outside of Alice Springs—very productive for various venomous snakes. Mostly brown snake variants, but also large mulga snakes. Mulgas are Australia's heaviest venomous snakes and, at over six feet, the second longest after the coastal taipan. These mud-colored monsters are the "prison-wing shot-callers" of the Australian venomous snake world. While not holding a spot on Australia's most-toxic hot list, they make up for it in sheer quantity of venom. Gargantuan venom yields of up to one gram of protein are delivered through thick fangs backed up by incredibly strong jaws. These can exert such pressure that the flesh being bitten is compressed, effectively driving the venom into deeper tissues than would be accomplished through the simple length of the fangs. We collected these snakes as our out-groups for the research we were conducting on the enigmatic pygmy mulga snakes, the smaller cousins that had only recently been discovered.

On our way north from Alice Springs, we came across Wycliffe Well in the Northern Territory—Australia's answer to Roswell, New Mexico. The local gas station had constructed quite an elaborate alien memorabilia montage that included six-foot-tall plastic aliens. It is a popular place for many simpleton-style travelers to indulge in their recreational chemical of choice and have an amusing time wandering around the otherworldly exhibits. From there it was up to the Devils Marbles, the large, almost perfectly round boulders which in Aboriginal tradition are believed to be the fossilized eggs of the Rainbow Serpent.

We continued north and into the zone targeted by my current research, which concentrated on the unique evolution of the animals that occupy the escarpment country. These huge crumbling ribbons run across the top end of Australia from Weipa and Mt. Isa in Queensland, through Litchfield National Park in the Northern Territory, and up into the Kimberley in Western Australia. They are much like coral reefs in the ocean: an oasis of life in a lifeless desert. I was interested in how the animals changed across the range to adapt to the subtle differences in geology and vegetation. In particular, we were after death adders and pygmy mulga snakes. As the curse of the cane toads was spreading quickly across northern Australia, we had to work fast to establish the biodiversity and flag unique species in need of special breeding programs. The research also had important implications regarding the relative efficiency of antivenom.

As we arrived in the tropics, my heart sank as I saw that the dark tide of cane toads had already swept into Kakadu, drowning all before it in foul poison and leaving only bleached bones behind. Even ten-foot crocodiles are not immune, as a single toad contains enough poison to kill them. This was not because cane toads are inordinately toxic; rather, the reptiles in Australia had evolved in the absence of toads. Thus they were naive to the toxins. Much in the same way as the indigenous tribes of the Americas, when the Conquistadors first arrived, died in waves from viruses and ailments that were common and typically non-lethal in Europe. The Australian tropics are plentiful in frogs, so many predatory animals will readily ingest a cane toad, not realizing the difference. Death adders and goannas in particular were hit hard. We found many rotting corpses, with an equally dead cane toad in the mouth.

I felt impotent, despairing rage at the arrogance of the Australian government. Acting against scientific advice, and with maddening stupidity, they had released a nocturnal toad that does not jump to feed upon a beetle that is active during the daytime and roosts six feet above the ground at night. At the time of introduction into north-eastern Australia in the 1930s, there was already sufficient evidence that cane toads rapidly become a pest, with Guam and Samoa having been similarly afflicted once these Darwinian monsters were introduced there. In the absence of any effective predators, the toads had rapidly become a plague in Australia. They were an absolute cancer to the country's fragile biodiversity. Even worse, the toads were adapting, becoming bigger and, more insidiously, longer-legged and faster with each successive generation, and had hit Kakadu years ahead of the most pessimistic of predictions.

Our primary targets at this section were the water snakes that inhabit Buffalo Creek. To catch these, we launched a small boat at night on the dropping tide as it exposed the dank mudflats. Shining flashlights, we would spot the water snakes in the shallows and net them out. But we were far from the only predators. This stretch of river was notorious for its plentiful "logs with teeth," or crocodiles, as other people called them. We had to be ever vigilant for eyeshine. But that did not stop a very large croc from sneaking up on us from below. The fish finder was insisting that a fifteen-foot barramundi was beneath us as the leviathan stealthily made its way under the boat, taking up most of the space between the base of the hull and the river bottom less than six feet below. I let go of the branch I was holding and allowed the current to take us away to safety. Neither of us were breathing for the next few minutes before the pendulum swung and Chris and I started hyperventilating. To make it even more stressful, we were only fifty feet from the boat ramp. We had to pull up into the shallows and get out into the water in order to be able to hook the boat to the trailer. We looked for telltale bubbles or V-shaped disturbances in the water as we worked as quickly and quietly as we could.

Next stop was the Barkly Tableland, hitting it right at the peak of a biblical-level insect plague. It was fantastic for the rest of the food chain, but not so good for us one fateful night. After a routine high-speed skid, leaving long streaks on the road, we shot out of the car to chase a mulga snake, forgetting to close the doors. Hundreds of flying stink bugs were attracted to the interior dome light and they were crawling through all the car's nooks and crannies for the rest of the evening. Out of reflex, when one would crawl up the neck, we would crush it like it was a fly, creating a fresh explosion of putrid chemicals. The next day the bugs all died in the heat, but the concentrated chemicals had been absorbed throughout the car, leaving a lingering putrid smell that persisted for the rest of the trip.

The following afternoon, Chris and I spotted rain clouds up near Cape Crawford, so we shot up that way, hoping the weather would bring out the massive local death adders—the same species that affected Chris's muscles so badly years before. Instead, we ran into a flash flood. While the rains were not near us, we were in the lowest part of the giant floodplain, with all the water coming our way. By the time we noticed this, the water was already past us. The road was only slightly elevated relative to the floodplain. It was not enough. Water was soon coming over the road as far as the eye could see—conditions treacherous for any vehicle but even more so for a two-wheel-drive station wagon. We turned around and ploughed our way back. The floodwater had topped the wheels by the time we got through the worst of it. Once we were safely out of the floods, we stuck near them all night long. The snakes were plentiful, including the biggest curl snakes I have ever seen. We caught several of these spastic snakes, which were almost three feet long. They have small beady yellow eyes on a black head that abruptly transitions to an ochre-colored body. These were a key species for the research and one that we had specifically targeted for this trip. While nothing was known about their venom, these snakes had a very bad vibe. It was a case of the personality being so malignant that I just knew the venom would tear a person apart.

The following day we headed east across the Barkly Tableland toward our main destination: Mt. Isa. We were only halfway across when the engine temperature began to climb to a dangerous level. Stopping to check things out, we noticed that the radiator was covered with a copious layer of dead insects. We scraped them off as best we could, but it was a like trying to shave a week's beard growth with a butter knife. We continued on, but ten miles later the engine temperature reached a critical level. One of the radiator hoses split, announcing itself with a geyser of superheated steam erupting from under the hood. We were now stuck in the middle of the desert in peak summer. The temperature was 120 degrees in the shade. Out in the sun, we felt like bugs on a sidewalk below the magnifying glass of a sadistic eight-year-old boy who has focused the sun into a death ray the likes of which Darth Vader would covet. In addition to the extreme hyperthermic risk to our own wellbeing, we also had a car full of very rare snakes. Luckily, thirty minutes earlier I had passed a white Kombi van. As we were on a road with no turn-offs, I knew he was still behind us somewhere. Fortune smiled on us—the guy turned out to be a mechanic from England who was driving around Australia for a few months. For travel money he had been fixing vehicles here and there. He had spare hoses with him and some sort of digestive spray that helped loosen the concrete layers of mummified insects. It took him no time at all to get us on our way again.

About halfway between Camooweal and Mt. Isa, we encountered a huge sandstorm. The approaching orange wall towered almost a mile high. We did a quick turnaround and accelerated away from it, but it consumed us despite the fact we were doing fifty miles an hour on the narrow, twisting road. We had to stop where we were and hope that a road train would not clean us up. It was such a fierce sight. The churning wind carried not just fine sand but larger, coarser pieces as well. This sand blasted the paint on the car, leaving countless tiny chips and dings. But, it had been purchased for reliability rather than stylish looks, and I was not terribly concerned about resale value since I knew that I would run it into the ground with all the field driving I planned on doing.

We arrived in Mt. Isa to act as co-hosts/guides/babysitters for a film shoot. Love at first sight just didn't happen between us and the host of the program, whom I shall refer to only by the code name "Big Guy," in that ironic manner in which one would call the big, muscular guy "Tiny." This referred not just to stature but also experience and self-awareness. He was an epitome of the Dunning–Kruger effect: the inverse relationship between knowing a little and knowing a lot, in regard to how one assesses one's own ability. The more one learns, the more one realizes how little one has learned. The know-it-alls are so often those who know the least. It can manifest itself in spectacular forms. In this case, it was a piece of white trash fronting an ambulance-chaser of a show built largely around him getting punished by all sorts of animals. He wasn't in on the joke that was the essence of the show and instead walked around emanating an overwhelming sense of entitlement, since, in his mind's eye, he was the Snake Saviour. Unsurprisingly, this could only end in tears.

It was readily apparent that he was as incompetent in snake handling as his tattoo master was in ink. His body art looked more like a child had drawn it on with a permanent pen. So we devised a plan to restrict his access to the snakes, partly for his own good. This plan was helpfully facilitated by an unseasonable cold snap that hit the night after we arrived, which meant that very few wild snakes were out at night. Just in case, I would quietly send Chris ahead of us in a separate vehicle to vacuum the roads for any available snakes; he only spotted a half dozen in total. The film crew was more than happy to fake scenes to get the shots they had written down on a clipboard. I would use the calmest snakes we had caught, but even so, I made sure that I was always the one who reached them first and "caught them" for the camera.

The last night of filming we were so thoroughly annoyed that we skipped having dinner with the crew. Right at nightfall, we came across the most gorgeous hatching black-headed python. A darling little thing with thick, clean cream bands alternating with narrow, crisp black bands, all offset by the ebony head. The film crew had been coveting a black-headed python all film shoot and had written an intricate scene involving one. Quite naturally, once the crew caught up with us we said nothing about it and drove the entire evening smirking to ourselves, as the snake remained unmolested in a bag we had specifically hidden under Big Guy's seat. Chris and I felt the host and crew were not worthy of such a magnificent creature.

Around 10 p.m., we were almost done with the entire shoot and were getting ready to head back for the ninety-mile drive across the desert to Mt. Isa when we came across a wild curl snake crossing the road. I bolted out and got it under control quite quickly. Just as I was back in the car loading it into one of the wooden travel containers for safe transport, over my radio-microphone I heard Big Guy exclaim, "There's another one!" I instantly had a sinking feeling in my chest. By the time I raced over, he seemingly had the curl snake under control, despite trying to pin it down on the gravel and grab it with a shaking hand from too far behind its head.

With haste, I divested him of the snake and secured it away. It was with great relief that we headed back to base camp. As we were unloading gear, while he stood to the side, stole oxygen and spewed out words of utter rubbish, I noticed a beer bottle slip from his grasp. I commented on this to Chris, since beers were usually held by this terminal alcoholic using the same sort of death-grip with which Gollum would grasp his "precious" in *The Lord of the Rings*.

Learning from Joe Slowinski's death, Chris and I had abstained from alcohol all week. This was as much for safety purposes as it was for public relations in the event of things going wrong. With the snakes securely locked away in containers for which only Chris and I had the keys, the crew let down their hair. In addition to the several beers Big Guy vaporized like a man who has been lost in the desert for a week, he fired up a joint and adeptly turned it into a cinder.

Around 2 a.m., when it was just Chris, Big Guy, and I still up talking, I noticed that he was walking with a very hunched-over and goblin-like gait. He also seemed to have something on his mind that he desperately wanted to impart to us. Eventually, he dropped a bombshell that painted the walls with my brains: this bloody idiot had been bitten by the curl snake that he'd tried to catch nearly six hours earlier. My initial reaction was stunned disbelief as my heart seized up like the pistons of an overheated engine in a rental car being trashed by a field herpetologist. When I stood up to get emergency help, in his delirium Big Guy even more inexplicably tried to stop me from doing that by wrapping his arms around my legs to prevent me from walking, causing me to fall to the ground. With alacrity and agility, Chris was upon us and grappled with this moron until I was able to extract myself and sprint from the room.

Upon returning, I found Big Guy prone, and Chris holding his head on his lap. Chris had a look of pure horror on his face, which no doubt mirrored the one on my own. My first thought had been that he was dead. Apparently once I left the room all the fight had left him, his eyes rolled up into his head and he gave a shudder before going limp. I was so frantic that it took me the longest thirty seconds of my life to find a pulse. Having no luck getting one off his wrist, I managed to find a weak one in his neck. My second thought was that no one would believe us that this fuckwit had been bitten earlier and didn't tell anyone. It was just too fantastic to accept. I just needed him to hang on long enough to regain consciousness and confirm the truth. He was then quite welcome to fuck off and die.

When help arrived, as expected none of the ambulance officers believed us that he was not only stupid enough to get bitten and not tell anyone, but that he followed this up with the consumption of alcohol and drugs. I was in an absolute fury, angrier than I had ever been in my life. Not because I cared about him—for all I cared the vultures could pick the bones from his rotting carcass if they could stomach the rancid meat. Rather, I was focused upon the potentially career-ending scandal that would certainly eventuate if he died without revealing the truth. The crew looked at Chris and me like we were devils incarnate. I didn't need a trumpet up my ass to know this was one seriously fucked-up situation.

At the hospital, things became even more farcical. One of the attending nurses insisted that snakebite had not occurred because the snake venom detection kit (SVDK) did not come up with a positive in any of the wells. She was convinced it was a much more mundane event: an ordinary drug overdose. Through gritted teeth, I patiently tried to explain to her that the so-called venom detection kit was misnamed. Rather, it should be called the "snake antivenom-match kit" since it did not "detect" venom per se. In fact, it was an immunological sandwich-assay—whichever well gave a positive result indicated which antivenom would cross-react with the venom. There conceivably could be venoms that did not elicit a positive result, and thus might not be neutralized at all by any available antivenom. Indeed, this was the entire focus of the research upon species precisely like the curl snake. The Dunning–Kruger effect reared its ugly head again: she was too completely convinced of her own knowledge to give any ground on this. The doctor who came in next was much more amenable to reason and agreed to administer the polyvalent antivenom since by now Big Guy was displaying the classic effects of severe neurotoxicity and myotoxicity. Polyvalent antivenom is made against a mixture of venoms from a wide taxonomical assortment of Australian elapid snakes. If anything was going to work it would be this.

By this time Big Guy was in a world of hurt, but no more than he deserved. The myotoxins were tearing his muscles apart, while the neurotoxins had combined with the marijuana and alcohol to produce a very bad trip. I can only imagine it was quite a bit like the time I watched *Scarface* while on hallucinogenic mushrooms. Bad idea. Great movie while stoned. A bloody nightmare on mushrooms. If this was his world, it was one of his making. I wanted the seven stages of hell in an inferno for him.

He was delirious and in no shape to recount the truth about how events had transpired. So Chris and I had no sympathy whatsoever for his plight. Three vials of polyvalent antivenom later he was conscious again and made a full confession. Any lingering doubts about someone actually being this stupid and still remembering to breathe were blown away when, at 9 a.m., he checked himself out of the hospital against doctor's orders and insisted that he was going to fly off with the crew.

By then Chris and I were already packed and ready to hit the road. My parting words to the shoot director were quite simply that if they allowed him on a plane against medical advice and anything adverse happened, they would quite rightfully be held liable under Queensland's draconian workplace safety laws. Further, they would be demonstrating their complete negligence if they didn't inform the airline of his medical condition, and no airline would let him fly without written doctor's clearance. He should not be out of the hospital, let alone subjecting his body to the additional physiological stress of flight. Whatever they wanted to do was their problem now; we had washed our hands of the whole affair and wanted nothing more than to put as much desert between him and us in as short a time as possible.

This was the worst experience of my years of fieldwork. The sheer irresponsibility of it was something I could not comprehend. I was just thanking my lucky stars for the opportunity to emerge from it quietly, reputation intact. Chris flipped on Danzig's song "Mother," our dark moods matching the lyrics perfectly. If Big Guy wanted to find hell, we'd gladly show him what it was like until he was bleeding. We sped out of there relishing the release from his odious presence.

As we drove out, our moods blackened to the darkest of night when we heard the news that the US had just invaded Iraq in an ill-advised military venture that had absolutely nothing to do with the horrid events of 9/11 but was destined to become history's greatest clarion call for disenfranchised Islamic young males to take up arms and become extremists.

The eventfulness of the trip was not yet over. As we cut down Queensland and into New South Wales on our way back to Melbourne, we were caught just outside Canberra in the giant bushfires that had erupted in the extreme summer heat. Several massive fires had merged into a super fire front that was hundreds of miles across. The heat was so intense that it was generating winds of sixty miles an hour, with burning embers landing over twelve miles ahead of the fire front. The heat and the wind combined to produce something that few people had heard of: fire tornados. The wind was swirling to such a degree and contained so much evaporated oil gas from the eucalyptus trees that it created a liquid fire vortex straight out of the furthest reaches of hell. It was so hot in some places that the road melted. We drove through Canberra right when things were at their worst, one of the last cars through before the highway to Melbourne was closed.

Back in the lab we investigated the genetics and venom profiles of the various death adder and mulga snake types that we had collected across the Top End. In collaboration with Wolfgang Wüster and David Williams, we compared the genetics to those of similar snakes found in New Guinea. Wolfgang is the most adept genetic taxonomist I have ever had the privilege of collaborating with, and David knows more about New Guinea snakes and their venoms than anyone in history. Much to our surprise, it transpired that there must have been multiple land bridges at various times during the evolutionary history. It was a fascinating genetic puzzle and unraveling it kept us happily occupied for many months. The death adders in the escarpment country were shown to be one wide-ranging species, while those on the intervening floodplains were another. The venom differences were also notable. While the escarpment death adder was classically neurotoxic in action, the floodplain species was consistently potently myotoxic, consistent with the aberrant effects that had occurred in the bite to Chris years ago. The death adder antivenom performed well against the neurotoxic effects, but less so against the myotoxic. In contrast, compared to other species of black snake, which are usually potently myotoxic (as occurred with my Butler's snakebites), the pygmy mulga snakes we collected were much more neurotoxic. The black snake antivenom, however, worked quite well against both effects.

After completing the research in Melbourne, I was back to Canberra for the "Science Meets Parliament" event. This is an annual gathering where scientists try to convince politicians to look further than the ends of their noses, generally stuck in a food trough. They need to see the stark wisdom that long-term investment in science brings economic returns far in excess of the amount spent—something that should readily pass the bleeding obvious test but was in fact frustratingly hard to get the politicians to accept. This is particularly the case in Australia, which fancies itself as the "clever country" but is still largely dominated by short-sighted live-in-the-moment populist cowboys who would rather, for example, fire up easily accessible coal than invest in solar technology. I did, however, enjoy the opportunity to catch up with former Minister for Science Peter McGauran and have a good laugh about the spotted black snake near-miss that occurred during my halcyon PhD days.

The Australian Research Council grants and fellowships for that year were also announced during this event. The idea was that the politicians would then be on hand to be praised for their generosity. However, they did not consider the flip side of awarding only 20 percent of the applicants. This is not to say that only these few were worthy of funding—the vast majority of the remaining 80 percent were of sufficient standard. Some applicants used a ministerial computer to check if they were on the success list, only to find out they were not. It was somewhere between tragedy and farce for them to be weeping in a parliamentarian's office. It was thus with mixed emotions that I found out that my fellowship application had been successful and that I had been awarded three years of funding and salary. This allowed me to take up the position of Deputy Director of the Australian Venom Research Unit. While I was filled with delight and enthusiasm, I had to temper my outward display out of respect for the unsuccessful majority wandering around with one-mile stares in their despairing eyes.

7

UNDERWATER WORLD

I quickly set up shop at the University of Melbourne, occupying my corner of the Australian Venom Research Unit. This lab
had been established by Australian icon Struan Sutherland after
he retired from the Commonwealth Serum Laboratories, where
he had headed up antivenom development and clinical consulting for many years. Not only was he responsible for key antivenoms, including the notoriously difficult-to-develop funnel-web
antivenom, but he was also the discoverer of the pressure-immobilization form of venom-emergency first aid. He was head of
Immunology Research at the Commonwealth Serum Laboratories
for twenty-eight years, until 1994, and then founder of the
Australian Venom Research Unit at the University of Melbourne.
He died in 2002, a year before I took up the position as Deputy
Director. It was quite an honor to work there. I set about organizing many dusty boxes. Contained within a group tucked away in
the far corner of a storage closet, I came across venoms that had

been stored in dried form for decades. I added these to my growing "to do" list and moved on.

Hungry for more discoveries and thirsting for new playgrounds, I started exploring a new research field—that of fish venoms. But first, I acquired a trio of dingoes: two that had a very small amount of dog in them (less than 10 percent), and one that was not only pure, but from the wild.

The first two were a male called Norton and a female I called Mera, while the pure female I named Cleo. Norton and Mera were strays, each having shown up at different properties and ending up at the local animal shelter. The tattoos on the inside of their ears indicated that they had been someone's pets at one time or another. Perhaps they had wandered off; perhaps they were abandoned. There was no way to tell. But they were obviously well accustomed to people and quickly settled into my house.

Cleo, however, was a different kettle of fish entirely. I got her from the Alice Springs animal shelter. She had been dropped off there by a family who lived on a large rural cattle station. Prior to that some Aboriginals had taken her out of a den when she was only a couple of months old. They had kept her for three weeks before abandoning her at the cattle station. The family had kept her for four weeks and she had seemed to be settling in well. That was until one fateful day when they let her out for a pee. Twenty minutes later they looked out to see that she had merrily slaughtered all fourteen of their chickens. So it was off to the shelter. My mate, Rex Neindorf, found out about her and rang me to see if I could take her, with full disclosure of her past. Based on my experience with the hybrids, I naively thought I had what it took to tame a wild pure dingo.

The difference between the hybrids and the pure dingo was astounding. Even such a little bit of dog had a remarkable effect upon temperament and behavior. They were very "doggy," and actually trainable. This is why in the wild the crosses become such nuisances, since they don't avoid contact with humans like pure dingoes. In contrast, Cleo was like a dog-shaped, orange-colored, recalcitrant jungle cat. She was very sweet when she wanted to be; other times she was more aloof than even the most socially dysfunctional house cat. Trying to scold her was like scolding a rock. She would stride majestically away, tail raised, and shoot me a brown eye. Collectively, however, they were an absolute delight to be around. I just had to make a few modifications around the property, starting with a ten-foot-tall metal-mesh fence with a three-foot in-slope at the top. The entire fence was then electrified to keep them from digging under or tearing the mesh off the posts. This gave my Mt. Dandenong property a certain fortress look.

By and large, the dingoes stayed contained, with only a few notable escapes. One instance in particular stands out. Cleo got out at 4 a.m. and started hunting the goats on the neighboring property. I let Mera out to act as an attractant for Cleo, since the two were usually inseparable. Their teamwork struck again, but not in the way I intended. Mera instantly shed her dogginess, let her dingo side take full control and started hunting too. Australia has a one-strike-and-you're-dead rule when it comes to dogs harassing livestock, and this would be strictly applied to captive dingoes. They were not trying to deliver crushing bites like a dog would; rather, they were trying to use the uniquely dingo, very long and blade-like canine incisors, to slice at the back of the legs and sever a tendon. When Cleo was distracted by one of the goats, I took the opportunity to do a superman fly-through-the-air and crash-tackle her, face-planting in the goat-crap-infused mud in the process. Cleo didn't struggle, as she knew the game was up, and the little traitor Mera instantly morphed back into her doggy alter ego.

Dingoes were not the only Australian mammal I really liked. Wombats also caught my eye. A midnight drive through the Acheron Forest was always productive for spotting them. I never tired of seeing these gentle giant herbivores, which were absolutely entrancing. On a clear full-moon night they could be easily seen just from the heavenly glow. "Under the Milky Way" by The Church was something played at least once (but usually several times) a night. It was a long stretch of gravel road. I would enter in from near Marysville, exiting near Warburton. I would spend around four hours there. This was one of my havens in the world. I always set up little emotional bolt-holes across the world; somewhere dear to bail to when things got a bit too much.

Acheron was an appropriate name for this spooky wonderland. Unlike reptiles, wombats were not seasonal, as these mammals always needed to eat copious amounts of plant matter, especially some of the roots that they considered tasty.

These furry tanks were like me: they would travel from point A to point B by going through things rather than around them. I could relate well to this method. This is not to say that they could not utterly fuck-up a person. Make no mistake about it, they are very dangerous animals. In captivity, babies are loveable little wonderbutts, such as Bean at my mate Stuart Parker's wildlife park in Ballarat. But as they age, they become solitary in the wild rather than forming herds like many other herbivores. Because of their eating habits, they have incredibly powerful jaws and strong teeth. I knew of a zookeeper who'd suffered a bite to the face that resulted in a crushed mandible and carved out cheekbone section; and another who had experienced a bite to the forearm with both the radius and ulna bones broken, with the radius being a compound fracture. So, I mostly viewed the wombats from the confines of my car, which did not seem to bother them nearly as much as if I stood near to them.

I could roll right up on them. Their blasé attitude to cars does not help wombats dying from motor vehicle hits.

I would also net out the rainbow trout introduced for sport fishery. I have nothing against the stocking of streams but not in fragile alpine forests that are refuges for rare frogs, as was certainly the case in the Acheron. Trout have to eat something and naive small frogs and their tadpoles were on the menu, so the fish had to go. I and the other members of the Trout Liberation People's Front would conduct midnight raids in the creeks, making them safe for frogs. The liberated trout were fed to my Mertens' water monitor lizards upon returning home.

One time I was out with my mate Kim Roelants, who was over from Belgium. Early in the drive we came across a big road-killed wombat. I told Kim that whenever we came across any sort of road-killed marsupial, we always checked for a joey in the pouch. So off he went into the darkness. In the low light, he relied on feel to find the pouch, sliding his hand around until he came to a moist opening. Only it was not the pouch. What had happened was that in the boiling summer heat, the insides had cooked. The body became swollen like a balloon as gas filled the intestines and stomach. Eventually the pressure became too great, and air escaped through a newly formed hole in the body wall weakened by decomposition. This was the hole that his hand entered before sliding into the mush that the innards had become. Truly disgusting stuff, even for my snakebite-damaged nose.

Later that night, after stopping and walking along with a half dozen live wombats, we came across more roadkill—another wombat. This one had been struck a glancing blow by a logging truck. Its neck was broken but its skull appeared intact—a perfect specimen for using in anatomy teaching. So we wrapped it up in a horse blanket. As unclotted blood was still coming out of its mouth, soaking through the blanket, we stuffed it, covered head first, into a three-gallon bucket. Upon reaching home, we put it under the house. The idea was to take it into the museum and preserve it by soaking it in 10 percent neutral buffered formalin.

The next day a big bushfire was burning in the Dandenong Ranges National Park, which is where my house in Kalorama was. Over the next two days, we were completely distracted by fire mitigation efforts, including plans for packing up the many animals. By the time we remembered the wombat, it had been under the house for three days. It was now a hairy sphere with its four legs sticking straight out. It was a methane-filled bomb that could go off with the slightest disturbance. We started driving to the same forest we had got it from, taking a new route to try and cut time, but we were soon lost. The smell was overpowering in the car as the wombat in the boot started leaking noxious gas. So we pulled over, gazing into the darkness, unable to make out where we were. All we could see was a dark slope. We assumed we were standing over some sort of ravine. So on "One, two, three!" we heaved the wombat, still cloaked in the blanket and bucket, on to the gentle slope. Late the next afternoon, a Tuesday, we went for another drive, retracing our steps. Pulling over, we discovered that the ravine was actually a slope down to the front lawn of a primary school, where we had left a child-sized object wrapped in a blanket with the head region tucked into a pale blue bucket. Oops—what must have gone through the head of the staff member who came across it first?

A much less amusing and more traumatic event occurred not long after this. My mate Tim Nias was diagnosed with leukemia, which meant that the white blood cells in his body were wiped out, just as had happened with my friend Jon from HIV/AIDS during undergraduate university. Also like him, this left Tim exposed to any bacteria that came along; however benign it might be under ordinary circumstances, it could be lethal now. As with Jon, the result was inevitable: bacteria made it into his bloodstream and killed him. Tim was one of the gentlest, funniest people I have ever had the privilege to call my friend. He was also a brilliant professional reptile keeper and absolutely gifted with venomous species in particular. He was a constant source of useful information to me. Whenever I had a tricky new animal, he was the first one I would ring for advice. His funeral in Sydney was attended by a broad spectrum of people in the reptile community, from private keepers through to the heads of the various zoological institutions. I know that Tim would have appreciated that after the service, while taking a walk in the grounds to clear my head, I caught a red-bellied black snake behind some bushes near his grave.

At this point I was maintaining a large collection of sea snakes at the Melbourne Aquarium, which had generously donated several of their very large tanks, and staff time, to help out with the research. I liked nothing more than going diving with the sea snakes with curator Diane Brandl and hand-feeding defrosted fish to the snakes. This was the most therapeutic thing I could do after Tim's funeral and it definitely soothed my soul. As part of the research, we tested a theory that my friend and collaborator Guido Westhoff had about sea snakes: that they were able to sense water currents, which would aid their fish hunting inside the coral rubble. Guido did a series of very intricate and precise experiments, consistent with his extreme German-ness, and was able to show that sea snakes indeed had this hidden ability. Therefore, we were able to not only discover new information about the venoms, but also about the behavior of these snakes I found so fascinating.

My new venom research effort at this time was on fish venoms, a relatively neglected field. Despite the vast number of venomous fish found in the seas, streams, and lakes of the world, research into them lagged far behind research into other venomous creatures. Indeed, more papers were published in an average year on snake venom than had ever been published on fish venoms. I quickly discovered why: fish venoms are slime-filled nightmares to examine. I concluded that the lack of progress in the field was not from lack of interest, but rather due to the inherent difficulty of working with the venoms. We tried a variety of techniques to separate the venom proteins from the slime without changing the chemical properties of the toxins. After much trial and error, we settled on a combination of ammonium precipitation at refrigeration temperatures, followed by acetone:methanol precipitation at freezing temperatures. Now that we had determined a way to purify the toxic proteins, it was simply a matter of getting more raw material to work with—the fun part!

My first collection destination was Trondheim fjord in northern Norway, where I linked up with the extremely affable marine researcher Jarle Mork from the Norwegian University of Science and Technology. I arrived in peak summer, which meant endless daytime in the land of the midnight sun—my favorite time of year in Norway.

We set out on one of the marine station's research vessels and put nets down very deep to reach the bottom, one mile below the surface. We were after a very strange fish called the havmus. A distant relative of sharks and stingrays, it had a dorsal spine over six inches long made of cartilage, not bone. Like stingrays, it was also armed with potent venom. My interest was piqued by a report in a Norwegian paper about a commercial fisherman stung by one while removing it from a net. In addition to the pain, he had wasting of the muscles in his leg and was partially crippled for weeks while the leg slowly healed.

Our first few benthic drags not only brought up abundant havmus, but also quite a large number of grenadier fish. Despite the grenadiers being a fish with a spine made of bone, and thus last sharing a common ancestor with havmus hundreds of millions of years ago, the two fish were extremely similar in appearance. This was a result of occupying the same ecological niche. Both are slow-moving predators at the bottom of the freezing cold fjords. Both have long, thin bodies that taper to a rat-like tail instead of the usual tailfin seen in fish. They look like an eel that has been decomposing on the bottom until only skeletal remains are left. They are certainly unappetizing to the eye, but both have a reputation for actually being extremely tasty, reflecting their crab-and-shrimp diet.

In two days of work, we were able to collect 110 havmus spines, which was an ample quantity to undertake the research, even taking into account the huge losses of material during the complex clean-up process. We also had a bonus come up in the nets—a halibut, quite rare in polar waters, though common in the more temperate waters further south. Global warming, with the increase in water temperatures, had extended their range.

In addition to the halibut, another unexpected but much more abundant animal came up in the nets: helmet jellyfish, usually denizens of temperate waters. I was astounded when we switched netting locations and ended up getting nothing but tons of them in the nets. This was an ecological and economic disaster unfolding before our eyes as it not only disrupted the food web, but would also ruin fisheries that had been sustainably managed for generations.

After the successful expedition to Trondheim, I stopped at the family island home on the Oslo fjord. My uncle Hans and I set out nets, this time not for venomous fish but for sea trout and flounder to eat. However, there is no escaping venom and we had several grey gurnard come up, a type of scorpionfish vital to the research. I knew all too well the power of scorpionfish venom. Before leaving Melbourne for this trip, I had been envenomated by a distantly related scorpionfish called the devilfish that my Norwegian cousin and I had collected while scuba diving in the waters at the southern tip of the Mornington Peninsula. I had been holding it in one hand while carefully raising the spines with the other hand. It gave a violent twist and kick with its tail. As it cartwheeled up into the air, knowing that an envenomation was inevitable, Haakon and I had just enough time to think the same thought: "I hope it's not me." But it was my forearm that it landed on, not Haakon's, driving its venom-tipped spikes into my flesh.

Within thirty seconds, I was in so much pain that I thought I would puke. As we were in the Melbourne Aquarium at the time, we sprinted for the bathroom and turned on the hot water tap. Fish venom toxins are very large proteins that are folded back on themselves to form a loose globular structure. This means that they denature very easily, unfolding just like egg white protein does when changing from clear to opaque white in a hot frying pan. That makes first aid for fish envenomations as simple as running hot water over the affected limb. But there is one major caveat, in that the pain from the venom is much greater than the pain of scalded flesh. So, the water temperature must be tested by someone not in pain, as it was by Haakon in this instance. There have actually been cases of people presenting to hospital with a foot that has been effectively cooked. The person has generally said that there was a reduction of pain but it still hurt, so they kept it in the water, not realizing that the toxin was long gone and they were instead boiling their foot. In extreme cases, this has led to the loss of the affected limb instead of what would otherwise have been an uneventful recovery. In my case, within about two minutes the pain stopped and life returned to normal. Well, as normal as it ever got for me.

A week later, Haakon and I were off to the Osprey Reef in the Coral Sea, with Freek Vonk flying over from the Netherlands to join us. This dive spot far offshore is famous for the very large resident moray eels. These intimidating fish occupy the same niche as sea snakes, but slice fish in half with their many razor-sharp teeth. These eels have even evolved a cutting second jaw just like the monsters in the *Alien* movie franchise. The upper gill arch has evolved to have long, sharp, grasping teeth. The eel latches on to the hapless fish with the forward teeth and then uses the gill arch teeth to deep-throat the chunks of flesh.

Deeper still, after all the red color has been bled out of the water, were giant stingrays. Immobile discs of doom, over six feet across. They are basically biological electricity receiving stations. Even in pitch-black, they can detect another animal from up to six feet away just from the electricity put out by all living things. They are able to accurately stab with the spine a grapefruit-sized object once it comes within three feet. The disc is made of pure muscle, thus providing the leverage to propel the spine with such force that it can pierce all the way through a sternum, the protective plate where the ribs meet and which protects some particularly vital organs in the chest. The venom-tipped spines may exceed eight inches in length and many species have two of them. They are strongly backward serrated, so that they grip and tear on the way out, leaving behind one of the most painful venoms of all. The tail moves with such force that it creates cavitation like a boat propeller, moving with a speed sufficient for a vacuum to be formed behind it, creating bubbles as air boils out of the water. This was definitely my kind of dive site.

Three days before the end of the trip, Haakon and I were

kicking back near the stern of the boat with Rush's song "Tom Sawyer" pumping out a mighty bass line when a pair of giant bat-like silhouettes swam by under the boat. They could be only one thing: manta rays. We quickly grabbed our flippers, snorkels, and masks, and dived in with abandon. We stayed above them, watching them glide below us. The gliders watching the gliding. Not a care in the world. How long we watched them, I will never know. However, once they dived down out of sight, I noticed we were now a very long way away from the boat, which was now a speck on the rim of the distant horizon. Our pride would not let us signal for help by waving both arms and making an X-shape. Not that they would have been looking for us, since in our excited haste we had slipped off the back of the boat without letting anyone know we were going. The distance didn't bother us since we were both good swimmers. The increased buoyancy afforded by salt water would facilitate a gliding freestyle stroke, so I knew my sea snake venom–destroyed shoulders would hold up. I was a bit concerned about the waves, though, as a strange, steady wind had sprung up while we were manta watching and white-capped waves were starting to build. In some cases, they were even forming surf breaks over the pinnacle reefs submerged ten feet below the surface.

About a third of the way back, I glanced down to see a fast movement. A sleek shape passed by at a high speed about fifty feet below us. Shark. About ten feet long. Big enough to do some serious damage. Not good. I wasn't able to see the species on the first pass, but I saw it quite clearly on the second. It was a black-tipped reef shark. Not usually a dangerous species, but this was an unusually large animal, and it seemed quite agitated, as evidenced by the arched back and jutting pectoral fins. On its third pass it came right at us, turning at about a thirty-foot distance. On the fourth pass, I dived down and expelled all the air from my lungs while hitting a long metal-singer note, causing the shark to turn sharply away. When it came at us again, Haakon and I dived together and did the same, with the same happy result.

I knew that we could not keep this up indefinitely, so I told Haakon to follow me and we sprinted towards the reef breakers. My plan was that we would hide within the washing machine surge, veiling ourselves in the visual and acoustic sensory cloaking. From there we would follow the fringing reef, which made a lazy curve toward the boat, although that would increase the distance we had to cover by at least half. It was a good plan, but the irregular currents and smashing waves put tremendous strain on the rotator cuffs in my shoulders that were permanently damaged from the tearing which occurred after my sea snake bite. It took us about ninety minutes to get as close as we could to the boat, leaving a five-hundred-foot stretch of open water between it and us. By now, we could see that the crew had not spotted us—there were no heads turned our way—so they had no reason to know we were being stalked by a shark. We paused, remained motionless and scanned the water. Seeing nothing, on "One, two, three!" we sprinted toward the boat, churning the water with our strokes. By the time we got there, my shoulders felt like they were being held together by fraying rope. But even Australian Olympic legend sprint swimmer Ian Thorpe could not have beaten me in that dash!

Once we dragged ourselves onto the back deck, we could see why our absence had not been noticed. Everyone was in the cockpit, staring at the radar pattern. Over the last three hours, a cyclone had sprung up. It had already been given a name: Cyclone Larry. We were in the open ocean—a peak area for cyclone development, where matters can go from calm to chaotic in a very short period of time, particularly in midsummer. The radar was showing an ominous spiral that already had the characteristic eerily calm center. The captain was on the radio with the mainland, which was reporting back that all boats had been put on high alert and ordered back to harbor, while vessels were banned from putting out to sea. Things were now critical, since we were a long way from the coast and much closer to the cyclone. Very close, in fact. The cyclone was predicted to head in exactly the same direction as we were. We secured all loose gear, untied from the mooring and fired up all engines full for the shore. The seas grew steadily as we made our way back. All joviality was gone and all that remained was a grim sort of silent camaraderie. We reached Port Douglas late in the evening and made some phone calls to change our flights. We were able to secure seats on the first flights out the next morning.

As we drove down towards Cairns Airport, the dawn light had an eerie tinge to it. The animal life was also unsettled. The birds were aflutter, the kangaroos were bounding around randomly, and there were abundant snakes on the road. After spotting a half dozen scrub pythons, we even came across a road-killed taipan, but did not have time to stop and take a DNA sample from it. We made it to the airport with only minutes to spare to catch one of the last flights out. Landing in Melbourne, we were greeted by images on the airport's television screens of overturned planes scattered around the runways of Cairns Airport, like the abandoned toys of a giant child. Other images showed massive damage to houses and other structures. It was rated as one of the biggest cyclones ever to have crossed the Australian coast and the biggest ever to have scored a direct hit on Cairns.

A few short months later, my phone rang. It was a reporter from *The Australian*, asking for a comment about Steve Irwin's death. After a moment of stunned silence, I asked for a few more details, since this was the first I had heard of it. I learned that he had been killed by a stingray while filming at Batt Reef, dying quickly. He had approached the stingray from above and behind while snorkeling, then dived down at it. I explained that this is exactly how a tiger shark would attack a stingray in order to kill it and feast on stingray flesh. Therefore, it would trigger an instant reflexive defense from the stingray.

I knew from my recent experiense with a scorpionfish that fish venoms are extremely painful. Irwin's last experience would have been one of sustained agony. The long barb of a giant stingray had penetrated his chest and pierced his heart. Thus, I continued, there were three possible causes of death: bleeding into the left lung; loss of blood volume; or the pressure of bleeding into the sac that surrounds the heart, which would, in effect, cause a slow-motion heart attack. The first scenario, drowning from blood in the lung, would be slower than the time he had actually taken to die and thus could be ruled out. Blood loss would be quick, as would a bleed into the pericardial sac. So it would be impossible to determine the exact cause without seeing the autopsy report or interviewing witnesses further.

Regardless, I stressed that stingrays are gentle, magnificent creatures whose venom is entirely defensive. They only sting when they feel they are in mortal danger. Their curiosity and innate intelligence make them fascinating animals to observe. The rule is: don't mess with them and don't invade their space. If they come up inquisitively and you remain calm, it can be one of those magical encounters. However, considering that Irwin was there to film the natural history of tiger sharks and how they predated upon stingrays, he would have known all too well the triggers that would cause a stingray to violently defend itself. One must take into account the basic premise of Irwin's filming strategy: to provoke an animal into a defensive display so that he could scream how dangerous it was. As I had taken more than my fair share of liberties with dangerous animals in the field, as part of my generally risk-infused daily behavior, I was in no position to judge him for what could be perceived as hubris.

It was therefore quite ironic when a bit later, while netting stingrays in Moreton Bay for the fish venom research, I was stung by a very large black stingray. The barb went straight into the meaty part of my thigh, penetrating until it hit bone. When the stingray yanked the barb out, it was completely defleshed of all the venom-delivering tissue, meaning a full shot of toxins had been deposited into the jagged wound. The pain was instantaneous and blinding, and the bleeding was profuse. I knew then what hell on earth Steve Irwin's last minutes must have been. My only conscious thought was that "stingray" was far too benign-sounding a name. Really, it should be called the "GivemeagunsoIcankillmyself-ray!"

As we were out in the ocean, hot water was not available so I had to just suffer through it and try to stem the bleeding. After three hours, the pain was bearable and the blood loss was down; by now it was just oozing out. The next concern, however, was infection. It was a puncture wound inoculated with bacteria, and the abundant fish slime left behind provided a fertile growth environment. I was on very strong broad-spectrum antibiotics for the next two weeks. My leg swelled up and got a bit red, but nothing significant. What was significant, though, was the havoc wreaked upon the muscles. The venom contained powerful myotoxins that caused muscle wasting over the next few weeks, and then, when I started working out with weights again, the calf muscle tore. It left me limping like a gang member for a couple of weeks.

8
PUFF THE MAGIC DRAGON

The tense interplay between my love for snakes and the life-threatening allergic reaction to their venom had reached a tipping point. My allergy had become steadily worse, and with it was a reasonable assumption that I was now deathly allergic to all elapid snakes' venoms. The real worry, however, was the untested potential for a cross-reactivity with other venomous snakes. As my previous research had shown, many of the components in elapid snake venoms were found in the venom of other types of venomous snakes. Not just the vipers, but also the various rear-fanged species I had been intensively researching. I was caught in the jaws of a life crisis. All my life had been about venomous snakes and I was now faced with that ending. It's said that any crisis is not just about danger but also about opportunity. The danger was clear and apparent; the opportunity, however, was obscured in the mist.

I packed up my field gear and headed out into the Western Australian desert with my mate and fellow researcher Chris

Clemente. I had no particular research target. Chris was doing biomechanical research on the running ability of varanid lizards, so if nothing else, I'd have a chance to play with one of my all-time favorite animals. We picked a Toyota Land Cruiser and trailer from the Property & Facilities Department at the University of Western Australia, loaded it with the usual copious amount of gear and headed out on the highway with Airbourne's song "Diamond in the Rough" banging out of the stereo. We cut inland, heading for Sandstone. Along the way, a beat-up ute passed us at high speed and then did a hairpin U-turn and skidded to a stop. A leather-faced man got out and flagged us down. It was my old friend Brian Bush. I had mentioned to him weeks earlier in an email that we would be heading this way, and the date we were leaving. He was waiting out in the desert to surprise us. Brian knows more about the desert of the Australian west and its scaly denizens than anyone: he is a wellspring of vital and obscure information. He also has an extremely demented sense of humor, so our trip was slightly delayed while he kept us laughing with twisted tales about his life well lived. The kind of life we should all aspire to!

Continuing on our way, we drove well into the night, camped for a few hours and then reached Sandstone the next day. As I was reversing the trailer off the dirt track, it jackknifed and there was a snapping sound like sticks being broken. Examining it, we discovered that one part of the hitch was being held together not with proper rustproof titanium bolts but cheap steel ones that had rusted almost all the way through. They could have snapped at any time, which would have caused a catastrophic accident if it had happened on the highway. Thoroughly unimpressed, we headed into the small town of Sandstone to see if there was anything we could scare up to fix the hitch. The local council handyman had a very well-stocked toolshed and was able to find the right parts to fit. He also knew a lot about the reptiles of the area and gave us some good local tips.

Back at the field site, we fixed the trailer, set up base camp, and got back to setting up the drift lines and funnel traps. As the name implied, the area had sandy soil with lots of small stones, so it was easy work to dig out a shallow trench to partially bury the over 330 feet of barrier fence that would guide the animals toward the traps set in pairs on either side every sixteen feet. The traps were a modification of the classic minnow trap, with a funnel pointing inward at either end. Any animal that went in would not be able to get out easily. We covered the traps with shade cloth since the sun was so brutally strong; by 7 a.m., it was already eighty-six degrees. During the midday heat, we just sat in the shade of our tents and suffered. It was well over 110 degrees, even in the shade. Our tents were warmer than they should have been, since they were exposed in the brutal sunshine. The only big tree in the area had a massive bull-ant nest under it. Lucky us. These ants are almost one inch long and extremely territorial. We only discovered them after pitching our tents in the shade and then being stung numerous times. Now it felt like we were camping on the surface of the sun. All we could do was remain motionless, other than drinking water and swatting off the countless flies that pestered us.

When it cooled down a bit in the afternoon, we got into the four-wheel drive and cruised the dirt tracks looking for life on the move. The shimmering heat caused the illusion of dark waves rippling along the track, so it took me a second to realize that one of them was solid, not made of ether. A goanna. A big one. We got out and quietly assembled the twenty-foot surf poles with nooses at the end. When the goanna stopped to look at me, I kept its attention while Chris did an exaggerated U-shaped walk to end up behind it. Slowly, quietly, he approached. Once he got close, he ducked down behind the scrub so that he was invisible. Hidden, he snuck forward, with only the surf pole visible, like an antenna from the world's largest cricket. With me giving instructions in a barely audible voice, he lowered the rod tip until the loop was level with the goanna's upright head and then brought it backward until it draped around its neck. Then, yanking back like he was setting the hook into a marlin, he drew the noose tight. I sprinted forward to dive onto the writhing five-foot-long lizard. My hands and forearms were protected by welding gloves from the large, sharp teeth of the carnivore and its long, recurved raptor-like talons. It was a desert yellow-spotted goanna—my first encounter with this species of varanid lizard. Its deep orange body was covered in yellow dots arranged in an intricate pattern, obviously an inspiration for Aboriginal dot painting. Its blue-grey head gaped and hissed at me while its long purple tongue flicked rapidly in and out in a decidedly snake-like manner.

As I gazed upon it, I noticed something curious that had escaped my attention previously, despite my long history with varanid lizards in the wild and in captivity: there was a bulge running the length of its lower jaw, located in the exact same spot as the bulge on the lower jaws of the closely related venomous beaded lizards and Gila monsters. I could not recall ever seeing anything in the literature about this, and my scientific radar lit up like a Christmas tree. Could there be something to this? I spoke to Chris and he readily agreed that a couple of the lizards could be made available for my research purposes if we caught enough. Which we did. Over the next week in Sandstone we came across thirty-five of this species, as well as several other species, including the short-tailed monitor lizard, the smallest varanid lizard in the world: five inches of adorable toughness. In its mind, it was a twenty-three-foot giant. Intriguingly enough, all the species we caught had the same bulge on the lower jaw.

Monitor lizards are the ultimate predators. Never before had I encountered such perfectly adapted animals. They are powerful yet agile, and quick to accelerate to hit a very fast top speed that they can maintain for considerable distances. They have a special bone in their throat, called the hyoid bone, that allows them to turbo-pump air into their lungs, like the old-fashioned bellows used to kick-start a fire. Even the scales are reinforced with a piece of bone inside each, which has earned them the nickname osteoderms—"bone-skins"—and is about the only reason these beautiful animals have not been wiped out globally by the skin trade. The distances they cover in a day are astounding. One morning, at 8 a.m., we drove out in search of a diesel can that had fallen off the Cruiser during the previous night's snake hunting. When we found it, we noticed the bird-like footprints of a large lizard, with the characteristic crescent moon marks left in the sand by the swinging tail. There were two sets of tracks, one approaching and one leaving. Clearly, this large lizard had detected the smell and had wandered over to examine the can. We drove across the red sand, following the parallel tracks. Two miles later, we came across the lizard. It was another large desert yellow-spotted goanna, one that had already walked four miles in the short time since daylight. We piled out and gave chase. Running at top speed, we chased it for 650 feet before diving on it just as it reached the entrance to its burrow. Chris and I were completely out of breath, but the lizard was obviously ready to go again.

We boxed up the lizards we needed, broke camp and headed out towards the town of Newman to ship them back to Perth. We drove at night to escape the heat, but this meant that instead of heat we had the pleasure of a kangaroo plague. These suicidally stupid macropods were in pandemic numbers, due to the artificial habitat improvement provided by the farmlands. Despite our best efforts, we hit six of them that night, including a couple with joeys. As we were deep in the desert far away from any wildlife rehabilitation facility, there was nothing to do but euthanize the joeys as quickly and humanely as possible before continuing on. It would have been cruel to leave them to die in the baking heat the next day and we certainly were not equipped to nurse them as we traveled. One large male kangaroo jumped into the side of the car at 2 a.m., giving us quite a scare as this face looking like the mask in the *Scream* movies suddenly appeared in the driver's-side window before disappearing under the rear tires and then causing the trailer wheels to leap off the ground as they went over the mangled kangaroo.

Reaching Newman near first light, we went to the airport and dropped off the boxes before continuing on toward the Pilbara. At the Auski Roadhouse, we had our first shower in a week. We were as feral as a homeless person on a forty-day bender. The water stung like acid as our millions of tiny cuts were opened and cleaned out. Our clothes were disgusting beyond belief. We had stopped changing clothes several days earlier, based on the weird guy-logic that we were so dirty anyway, putting clean clothes on without washing ourselves first was just pointless. We just stopped wearing underwear, going commando to keep everything ventilated. Few tourists would ever have encountered two more filthy individuals walking through the desert with long fishing rods, looking like very lost dementia patients. We also removed carcasses of road-killed kangaroos from the road so that goannas or wedge-tailed eagles would not be car-struck while getting a free meal. But first, obeying the inscrutable exhortations of our demented senses of humor, we would draw a thick chalk outline around the body as though it were a crime scene, thus confusing any tourists even more.

The Pilbara has a certain prehistoric, harsh beauty to it. This oasis in the middle of the desert is carved out of black ironstone. During the day, we relished the novelty of swimming in the water-holes while still being in the boiling desert. The presence of the ironstone made night-time road cruising for snakes a low-yield venture. The ironstone's density, which allowed the rock to retain heat longer than the road, made it a more attractive option to nocturnal snakes looking to warm up before going out hunting. We did come across a few Pilbara death adders away from the ironstone and in the sand dunes. Selection pressures exerted by predation resulted in the evolution of solid orange sand camouflage instead of black and burgundy bands.

After another half dozen days in the desert, we stopped at Shark Bay on the coast for a bit of beach time. While drifting on my back in the waters, I reflected upon the name. Sure there were plenty of sharks around, but so what? Every bay has sharks in it. There was nothing particularly special or noteworthy about this one to warrant such a name. What was special about it, however, were the stromatolites. Ten thousand years ago cyanobacteria (blue-green algae) began constructing these unique structures that are analogous to the earliest form of life on earth. The oldest such structures are 3.5 billion years old, and fossilized versions can be found in Marble Bay, Western Australia. It was only in 1956 that living versions were found in Hamelin Pool at Shark Bay, giving stromatolites the longest continual biological lineage. Perhaps this area should be renamed Stromatolite Bay to reflect its true uniqueness! While stromatolites are incredibly interesting from a biological perspective, from an adrenalin junkie perspective they are incredibly boring. Picture a bunch of black tar-like lumps, with the occasional bubble coming from them. That's it. Nothing more to describe about their day-to-day, month-to-month, or even year-to-year activity.

We returned the vehicle to UWA completely thrashed. We had committed slow-death vehicular manslaughter. Every reflector was broken, the paint was pockmarked with gravel chips, the transmission sounded like popcorn, all the panels were dented, and two of the windows were broken. This was pretty typical of how we treated rental vehicles but our innovation this time was to duct-tape to the bullbar a ram's skull that we had named Wally. The vibrations of the next nine thousand miles of rough driving had resulted in it wearing two holes in the hood of the car—not just in the paint but almost all the way through the metal itself. But to us it was worth it, since the trip had been such a success and I had found a potentially exciting new path of research.

Upon my return to Melbourne, I set about in earnest to examine the monitor lizards for the possibility that they might in fact be venomous. Dissections of preserved specimens revealed a large macaroni-shaped hollow gland. Construction of cDNA libraries, as I had been doing with the venomous snakes, revealed that not only was there a diversity of proteins being secreted by the glands, but that some of them were in fact of the same type as classic snake toxins. I had goosebumps crawling across my body when I first examined this data; the purest and most intense feeling of elation. It suggested that not only were there more venomous lizards out there than previously recognized, but that snakes arose from these venomous lizards. This explained the curious frothy liquid coming out of the alligator lizard's mouth as it chewed on my classmate back in third grade, and also why he bled so profusely. It also explained why I bled so much when the water monitor lizard clamped down on my thumb that time at Arun's, when I was a university undergraduate.

This was as radical a reshaping of the view of reptile venom evolution as could be imagined. At the same time as I made this discovery, my good friend Nicolas Vidal made a startling discovery of his own. He had used new genes to assemble the first robust reconstruction of the organismal evolutionary history of the reptile lineages, placing snakes with anguimorpha lizards (a diverse assemblage including Gila monsters, monitor lizards, and alligator lizards) and iguania lizards as part of a new assembly. While snakes being close to anguimorph lizards was not a new idea, the closeness with iguanian lizards was very surprising. We examined the lizards and found that they also had protein-secreting glands. Freek Vonk did a sterling job on the histology for his Bachelor's degree research project and this was an invaluable contribution to the overall research project. While these glands were in a much more primitive stage than those of snakes or anguimorph lizards, they secreted some of the same components. Nicolas, Freek, Holger Scheib, and I combined forces and submitted the manuscript to *Nature*, the most elite of scientific journals. In this paper we gave the new venom clade the name Toxicofera.

While the paper was under review, I was awarded a travel grant from the Australian Academy of Science, and it was back to Europe for the northern hemisphere summer, which coincidently was right when Melbourne was at its most suicide-inducingly grey and depressing. At the Swiss Institute of Bioinformatics (SIB) in Geneva, Holger and I got to work on a variety of computational projects. Our basic approach was to have a nice breakfast at a sidewalk café, get into the lab around 10 a.m., load a massive dataset on to the computer cluster, and then go for long walks around Geneva. I particularly enjoyed our afternoon walks through the old town, with its historic architecture and mysterious doorways. We would then stroll along the lakeshore, with the spray from the fountain forming fleeting rainbows in the summer sunshine. Geneva is a surreal world. It is like just the top 1 percent of a mega-city has been relocated as a beautiful lakefront small city. The sidewalks were even designed to glitter at night, as if they contained millions of tiny diamonds.

I nipped up to Zurich for a few days to milk the viper collection of my friend Marc Jaeger. He had the most impressive collection of tree vipers from all over the globe, including eyelash vipers from Panama and bush vipers from Kenya. Unfortunately, not long before, he had had a total system failure of the misting system, resulting in a number of the cages flooding overnight and a massive number of deaths. While this was a huge loss of precious life, it was also an opportunity to further study the genetics and morphology of these highly specialized snakes. I was going to take them down to Geneva to deposit into the local museum.

It soon became apparent that we did not have enough room in the white styrofoam boxes for the snakes as well as the icepacks, so I opted to leave out the icepacks so that I could fit in all the snakes. My delusional hope was that they would keep each other cold during the train travel back. Unsurprisingly, this did not work. We were not long past Solothurn when the stench of rotting snake started wafting through the first-class carriage. I had stored the boxes on the upper storage racks and concealed them with bags and jackets, so it was impossible for anyone else to tell where this evil smell was coming from. By the time we were halfway to Geneva the carriage was otherwise empty of passengers. It was another situation where my sense of smell, having been destroyed by the two Butler's snake envenomations during my PhD days, came in handy.

Once at the museum in the lab of my friend Andreas Schmidt, we got down to sorting and cataloging the snakes and then storing them in formalin preservative. We were halfway through when I took a break to check my emails. To my great delight, there was

An unusually patterned reticulated python in Singapore. It was very rare, lacking almost all black pigment, which was instead replaced by the most beautiful gold color. However, as the blood on my hand shows, it was just as nasty as any normal one!

Inset: An olive sea snake I captured on a 130-foot scuba dive. It was the first one of this species I ever caught.

In Weipa, Australia, a bin full of spine-bellied sea snakes, the first sea snakes I ever caught.

Playing stick with my dingo, Norton, at my mountain home in the Dandenong Ranges, Victoria. Dingo saliva gives sticks a very interesting flavor.

With a huge male desert spotted monitor in the Pilbara, Western Australia. He is standing in his threat position and is pretty intimidating! This was the first lizard I ever found venom in, opening a whole new field of research.

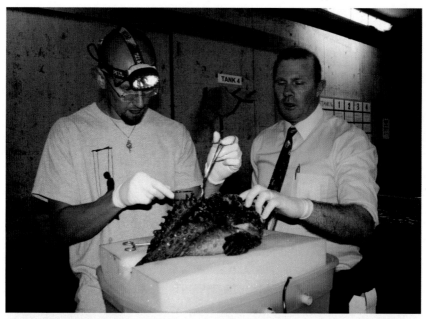

Milking a stonefish with Dr. Rob Jones at the Melbourne Aquarium. When the fish is gently squeezed, the glands on either side of each sharp spine squirt out their venom, allowing us to collect this death-inducing chemical cocktail.

Filming Komodo dragons on Rinca Island for a BBC *Natural World* special on my research. The stick in my hand is supposed to keep them away, but I suspected it would be more useful for the Komodo dragon to use as a toothpick. It was, of course, scientifically necessary for me to be shirtless on camera.

Left: Taking "giving the finger" to a whole new level with a large scorpion I caught in India while filming *Asia's Deadliest Snakes* for National Geographic TV.

Below: Milking a thirteen-foot Malaysian king cobra as part of a project looking at changes in venom across their vast Asian range.

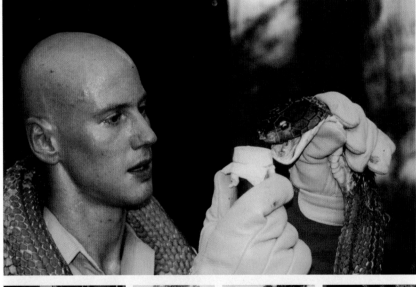

I was not a normal child.

With a lace monitor, one of the study subjects for the research showing that lizard venom is much more widespread than previously believed.

With boxes of Thai Red Cross snake antivenom ahead of fieldwork in Asia: the same sort of antivenom that might have been able to save the life of my friend Joe Slowinski, who was killed by a krait bite in Myanmar. There was no antivenom or doctor on that ill-fated expedition.

With a local snake charmer and my mate Sean McCarthy while at the Zamzama gas fields in Pakistan. One of the friendly locals; as opposed to the type who sprayed Taliban graffiti on the side of buildings and later on blew up our safe house.

Catching feral chameleons in the mountain forests outside of Honolulu, Hawaii—what every couple does on their honeymoon. Chip Cochran is in the background looking for more. My T-shirt perfectly conveys my love of metal music and my ability to be easily distracted.

A rough grenadier that came up in the nets we set 2,380 feet below the surface of the fjord in Trondheim, Norway, while surveying what sort of venomous life lived in their freezing depths.

Top left: At the Leiden University Medical Center in the Netherlands, examining a Komodo dragon head from a medium-size female on loan from the Berlin Museum. The plan was to use magnetic resonance imaging (MRI) to determine if these iconic lizards actually have venom glands. The BBC filmed this historic event for a *Natural World* special.

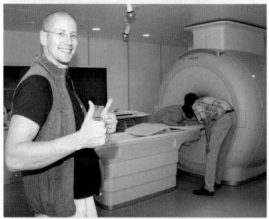

Left: Near midnight we are loading a Komodo dragon head into a clinical scanner at the Leiden University Medical Center. The hospital administrators gave us special permission to come in after-hours and use the equipment.

Conducting research using magnetic resonance imaging (MRI) machines at the Leiden University Medical Center in the Netherlands. My ear-to-ear grin as we get the results from the first MRI scan ever of a Komodo dragon's head, showing the venom glands that had been overlooked all this time.

In Parliament House, Canberra, receiving the Australian Research Council's Queen Elizabeth II Fellowship, one of only sixteen awarded that year and one of only two in biological sciences. Ollie the olive python caused quite a stir with security.

Ollie the python with minister Julie Bishop while in the real snake-pit: Federal Parliament. Ollie was so well-behaved that we fed him a huge rabbit upon returning to Melbourne.

Victorian State Premier John Brumby looking less than thrilled as Ollie the olive python smears his way across Brumby's back, leaving a trail of freshly digested rabbit across the suit. Ollie had defecated in his bag en route to Victorian State Parliament where I received the Victoria Fellowship. (It was the same rabbit we fed to him after he was so well-behaved with Julie Bishop ten days earlier. Oops.)

Reading the morning paper with Otto, my favorite spiny-tailed monitor lizard on my head while taking a break from fieldwork at my mountain house in the Dandenong Ranges.

Three desert-loving creatures: me with a perentie monitor lizard from Alice Springs and Norton, my long-legged desert dingo.

A nearly six-foot-long western diamondback rattlesnake I caught in New Mexico after wading across a section of the Rio Grande river. In addition to having an unusual reddish color, this population also had an unusual venom that was not well neutralized by any of the available antivenom.

The laboratory freezers are always packed near bursting with a weird combination of frozen specimens and test tubes.

Catching river sharks in the Gulf of Carpentaria at Weipa, Australia, with my wife, Kristina, who is always up for a new field adventure.

Above: A mulga snake I caught in Kununurra, Western Australia, just before the cane toads spread to that pristine region, wiping out these magnificent snakes. The venom samples we collected may be some of the last ever available to work on.

Right: An over-thirteen-foot-long king cobra in Malaysia. These highly intelligent snakes are very attuned to movement, so it was fixated upon my hand as I traced slow patterns in the air.

Left: The fear-and-confusion-inducing road directions in the cloud forest of central Mexico.

Below: Scorpions are in plague proportions in Mexico during the three- to four-month-long wet season, with up to 500,000 people stung and 150,000 of those requiring anti-venom. On our typical night hunting, more than one hundred scorpions could be easily collected.

A blue sea krait in the Andaman Islands of India. I was able to catch it despite being barely able to stand at this point. Unbeknownst to me, my back was injured much more severely than I thought. It was actually broken, and not long after this I would check into the hospital, not to emerge again until four months later with a spine rebuilt out of metal like the X-Men character Wolverine.

Three very toothy grins: with Kristina and a young alligator at my mate Marc Jaegar's facility in Zurich, Switzerland. It was times like this early in our courtship that let me know that she was the one for me.

an email from *Nature*, informing me that our paper on the single, early evolution of reptile venom had been accepted. I couldn't breathe. The potential impact on my career prospects was quite staggering. Getting a paper into *Nature* is like winning a gold medal at the Olympics. While it does not mean "set for life," it gives a massive momentum boost that can be used to keep a career moving forward for a very long time. I kept the news to myself for the rest of the day, avoiding Holger entirely. I didn't trust myself to be able to keep a straight face. I had come up with a cunning plan for how to tell him. The next day, I was scheduled to give a seminar at the Swiss Institute of Bioinformatics about the lizard venom research, on which Holger and I were collaborating. I started my lecture by dropping the news about the paper being accepted. Holger's face turned scarlet with the shock to his system, then white, then red again and finally into a wax dummy blankness. He did not hear another word of my entire lecture. That night, we celebrated properly in glittering Geneva. It was the best day of our academic lives.

Back in Australia, the floodgates opened. I received in rapid succession a number of scientific awards based on the success of the *Nature* paper and the slew of follow-up papers. This included landing a new fellowship from the Australian Research Council. I was awarded the coveted Queen Elizabeth II (QEII) Fellowship—one of only sixteen given out that year and one of only two for Biological Sciences. The ARC wanted me at the grants announcement as the face of their fellowship scheme. There was one catch: Peter Høj, the head of the ARC, wanted me to bring a python into parliament, the seat of the Australian Federal Government. Naturally, I accepted this quite reasonable request and rolled up to Canberra with an over-ten-foot-long olive python named Ollie, who belonged to my mate Chris Hay. Despite prior arrangements, the parliament security services still had a total freak-out about the python. They even X-rayed it before letting it in. They were so distracted and concerned about it that they didn't notice that the drill I was using for the screws that held on the wooden box lid was also a fully functional nail gun. So there I was in parliament with not only a very large python but also a lethal weapon. Naturally, I kept this to myself, and resisted the urge to shoot the science-budget-slashing Prime Minister John Howard repeatedly in the forehead, even though the ghosts of Australian native wildlife were telling me to do it.

However, it was Ollie that was in the true snake pit. Career politicians give off a very strange vibe. The Minister for Science at the time, Julie Bishop, freaked me out in particular. Once the cameras started rolling and she was posing with the python, she did not blink once during the ten minutes the cameras were focused on her. This, of course, prevents any one-eye-half-closed photos that make a person look like they are drunk, so fair call there, but the level of self-control this takes is eerie and not natural. She was as unblinking as the snake wrapped around her torso. I concluded that the only rational explanation was that she must be one of the Lizard Illuminati. She was definitely a kitten-eating evil reptilian humanoid from another planet. When *Men in Black IV* comes out, I expect to see her up on one of the big screens back at central office where they showcase aliens who have integrated themselves as prominent public figures.

Many other politicians and their staff enjoyed posing with the always-calm Ollie, to the point that he crowd-surfed around for over an hour while I stood back and watched with amusement. I then gave my presentation with Ollie twisting slowly around me, finishing with most of him balancing on my shoulders with the rest on the top of my head.

I was so happy with Ollie's superstar performance that on returning to Melbourne we fed him a very large rabbit. A week later I received a phone call reminding me about my invitation to the Victorian State Parliament to receive another award and a nice big check to use for additional travel. So we packed Ollie back up and headed for the reception in Queen's Hall. As we undid the lid from the box, a familiar smell wafted up, a smell that could mean only one thing: Ollie had crapped in the bag on the drive over and was now liberally coated with digested rabbit. I frantically and unsuccessfully tried to wipe him off as he came out. Which worked okay for the first three feet, but not the rest of him.

The Victorian Premier John Brumby wanted to pose with him, having seen the press of Ollie in Federal Parliament, so he put out his arms. Ollie slid across them and then wrapped himself around Brumby's back. While the inevitable accompanying skid mark was hidden against the dark of Brumby's suit, Ollie's passing did leave quite a noticeable stench. Perhaps Ollie had cunningly planned this, one of the native wildlife getting their own back, since Brumby was as much an enemy of the environment as could be imagined. With friends like Ollie, who needs enemas?! I could see Tania Smorgon, the sponsor of the award, smirking a bit as she watched the scene unfold. I got the distinct feeling she wasn't Brumby's biggest fan. To her eternal credit, she rolled her sleeves back and posed with Ollie. Her unflappability may have had something to do with the fact that her two young sons had a pet python at home. Showing up Brumby was worth taking one for the team.

As they could not take the check back, I was off to Leiden University in the Netherlands the next week to examine the venom gland diversity of snakes and lizards using magnetic resonance imaging (MRI) with my friends and collaborators Rob Nabuurs,

Wouter Teeuwisse, Louise van der Weerd, and Thijs van Osch. MRI is the non-invasive technique used for a diversity of medical exams such as scanning the brain for tumors. But we were going to use it for a rather different purpose: scanning the preserved head of a Komodo dragon I had on scientific loan from the Berlin Museum.

With the pragmatic approach to matters so very characteristic of the Dutch, we had approval from the Leiden University Medical Center to come in after hours and use the clinical scanners. Starting at 6 p.m., we worked until 3 a.m. to get the parameters right in order to get the highest resolution scans possible. It became readily apparent that the gland was not constructed of a single compartment with a single duct leading up to the teeth, but was actually extremely complex. There were six separate compartments, each with its own duct leading up to terminate between successive blade-like serrated teeth. It was clear that not only did Komodo dragons have venom glands, but that these glands were large and extremely intricate structures. Scanning Gila monster and beaded lizard heads showed that the Komodo dragon, and the variety of other varanid lizards we examined, had similarities yet also differences in their gland structures. Combined with our chemical analysis of Komodo dragon venom, we were quite confident in our conclusion that not only was the Komodo dragon the largest living venomous animal, but also that its extinct close relative Megalania was, at twenty-three feet, the largest venomous animal ever to have lived.

Of course, any laboratory finding must be backed up with studies of the natural history of the animals themselves and the role venom plays in predation. So I was off to Komodo Island with an Animal Planet film crew. Joining us was Kevin Grevioux, the creative genius behind the superb werewolf versus vampire

Underworld movie franchise. While I am an adrenaline junkie, I am also a complete geek for this movie genre, so I considered it a great privilege to be on camera with Kevin. We flew first through Bali, Indonesia, and then onwards to Flores. Indonesia was very much the modern day crashing upon the rock of antiquity. Trash was strewn everywhere, with sea garbage washing up on otherwise pristine shores as the modern intruded upon the ancient.

As it was the dry season, we didn't have to worry too much about mosquitoes transmitting diseases such as malaria or dengue, but the heat was extreme. At the Ranger Station on Rinca Island we filmed the large Komodo dragons that were hanging around waiting for an opportunistic feed on leftover food. Not long after, we left the station and headed up the large hill nearby, where we came across a large Komodo dragon on the path in front of us. Then we noticed there was one behind us and two more to our right. Our large, slow-moving group had attracted considerable attention from these formidable predators. The two largest started coming towards us with a deliberate gait that reminded me a little too much of a shark on an approach trajectory, so we dropped the gear to lighten our load and hasten our departure. We headed straight up the hill for about 750 feet before we turned to check behind us. The three largest dragons were still determinedly following us. Kevin asked, "What happens if they catch up with us?" to which I gave the concise reply, "If they do, it will be like the werewolves feeding in your movie. . . . All that will be left of you is a stain on the grass!" Our speed instantly increased. We walked for nearly half a mile before the last of the dragons dropped off the hunt. Using binoculars, I could see that all three had taken up a strategic position in the shade of the trees near the path to await our return. This displayed a level of predatory foresight that was chilling and amazing at the same time.

In order to avoid the awaiting dragons, we took a different path down via the steep side of the mountain, which was devoid of thick vegetation that could conceal a large Komodo dragon. From there we traversed the island over to the watering holes. Six hundred and fifty feet before we got to them, I could smell the stench of sewage-filled water on slow boil in the tropical heat. Komodo dragons had long been thought to use weaponized bacteria to kill their prey. But there is a simple saying in science: extraordinary claims require extraordinary evidence. The use of killer bacteria as a hunting weapon was as extraordinary a claim as had ever been put forward. The evidence for it was startlingly lacking. Despite this, it had long been a staple item staged in nature documentaries, resulting in a penetration into the popular consciousness far exceeding any other "fact" about these magnificent animals.

In fact, the primary role of toxic bacteria in debilitating and ultimately killing Komodo dragon prey was entirely speculative. Popular reports of bacteria festering on chunks of flesh remaining in the serrated teeth, or on gore around the head area, is in conflict with the documented behavior of Komodo dragons, who lick their lips and scrape their mouths to remove any gore remaining from their meal. This behavior result in the known characteristic of healthy Komodo dragons having clean mouths, pink gums, and white teeth that lack retained prey item flesh. The bacterial hypothesis is also in conflict with the preference of Komodo dragons for fresh prey over rotting corpses. The voluminous references in regard to bacteria ultimately rely on a book published in 1981 by Walter Auffenberg. However, when this primary source is examined, it is merely stated as a tenuous hypothesis that the gum and blood "combination may have some significance as a bacterial culture medium." Despite a conspicuous lack of support from quantitative data, this single statement is later extrapolated by

Auffenberg to "unsuccessful attacks often lead to a bacteraemia in the prey that eventually results in their becoming food for the ora population." In other works, this is repeated as dogma. Thus, without any supporting evidence, a strong association was put forward between bite wounds and severe wound sepsis and septicemia. In short: scientific voodoo.

The one study that later looked at the contents of Komodo dragon mouths found the same sort of bacteria expected to be found in the mouth of any reptile or the typical gut contents of a mammal. The one quasi-pathogenic species (*Pasteurella multocida*) was found in only two out of the twenty-six wild dragons. Further, no single species of bacteria was found in all wild dragons. The variability was quite profound. This points toward transient, environmental sources for any bacteria present in the dragons' mouths—not a scenario that would be evolutionarily favored for use as a weapon. Rather, a typical scenario for any carnivore.

Komodo dragons evolved in Australia, not Indonesia, and are by no means the biggest to have roamed. At least two (now extinct) larger varanids existed to predate on megafauna. The second largest, at over fifteen feet, radiated to the island of Timor, while the Komodo dragon radiated to Flores and other nearby islands. The largest species, *Varanus (Megalania) priscus*, remained in Australia and reached well over twenty-three feet. The situation today is that the Komodo dragons have three mammalian potential prey choices: deer, pig, and water buffalo. Deer and pigs are within the natural prey size (ninety to one hundred pounds) that Komodo dragons would have evolved to eat in their native Australia, while the water buffalo are dramatically larger. Field observations by others and myself showed that when Komodo dragons attack such prey items as pigs or deer, 70 percent die in the first thirty minutes. These deaths are from the profound mechanical damage wrought by the shark-like teeth, with the animals dying from severe blood loss. A young boy was killed in this way at one of our field sites on Rinca Island. A large Komodo dragon attacked him from behind while he was squatting and defecating. The initial attack sliced his femoral artery all the way through, resulting in several massive blood spurts, each to a shorter distance than the one before as the boy's blood pressure dropped substantially due to the loss of blood volume. He was dead in less than two minutes.

Intriguingly, another 20 percent of attacked pigs or deer continue to bleed profusely and without clotting, and die of blood loss within three hours. During this time, the pigs or deer appeared sedated, walking slowly and with an unsteady gait, and go into shock long before they should from blood loss alone. Ironically, Auffenberg himself recorded similar observations in his book, thus inadvertently providing the first evidence for the dragon's use of venom. Our laboratory work had identified in their venom toxins that prevent blood clotting while also precipitously lowering blood pressure. These chemicals are consistent with the signs and symptoms observed in the field.

The outcome of attacks on water buffalo was dramatically different. The buffalo would get away every time, but with deep wounds to the legs. They ran to safety, ending up standing in feces-filled waterholes, thus facilitating dramatic infections—not from the dragon's mouth, but rather from an environmental source. A deep wound in such water is a perfect scenario for the flourishing of bacteria, particularly the nasty anaerobic types. Thus, the sampling of Komodo mouths that purported to show them harboring pathogenic bacteria neglected to sample the real source of any infection to the water buffalo: the waterholes. It has been an artificial scenario all along that has nothing to do with the evolution of the predatory ecology of Komodo dragons.

Water buffalo are an introduced, feral animal on the islands.

Any interaction is unnatural, even if Komodo dragons had evolved on the islands, rather than being Australian in origin. In their native habitat in south-eastern Asia, water buffalo live in large, clean, free-flowing marshes. Thus any feces are carried away, diluted, and broken down. However, on the rocky islands they share with Komodo dragons, the only water sources are backyard swimming pool–sized stagnant watering holes that rapidly become choked with sewage. Ultimately, these disgusting waterholes are the source of any infection the water buffalo suffer after a Komodo bite. Therefore, we concluded that any infections that do occur are inadvertent, and that the idea of using bacteria as a weapon has as much to do with reality as the sun orbiting a flat earth.

Returning to Flores at night, a week after we left, we had to unload the mountain of gear from the boat. As I was stepping onto the slippery steps exposed by the low tide, a sneaker wave hit the boat sideways, sending me tumbling off the boat. I flung up one hand to grab a rope I could just make out in the flat light of dusk, caught it, but then, swinging sideways, smashed against the pier with my left side, with my knee bearing the brunt of the impact before I went into the dank harbor water. The crew pulled me out and laid me flat on the pier walkway. A crowd quickly gathered, giving me a suffocating feeling as I writhed in pain. Kevin channeled his character Raze from *Underworld*. Baring his fangs, he roared for them to stand back and give me air. Even if the locals did not know what he was saying, they got the point from the intimidating posture of three hundred pounds of ebony-skin-covered muscle.

I was transported to the local medical facility, which consisted of a single concrete room with bare light bulbs illuminating the geckos running across the paint, which was peeling like a skin-cancer-inducing sunburn. It was only slightly more hygienic than the pollution-filled water from which I had just been pulled. My knee looked like hamburger meat, and the side order of chips was the abundant bits of barnacle shell sticking out of my flesh in random directions. The young doctor was partially blind, due to a cataract that caused one eye to look like that of a dead fish. He recognized me from television and seemed more interested in posing for pictures with me than attending to my considerable injury. He poked half-heartedly around the wound before sewing it using the needlecraft of a drunken housewife. Even though he used half the number of stitches it required, this took my career tally to four hundred stitches.

We flew out the next morning to Bali and checked into a hotel in Ubud. By evening, my knee was hot and inflamed. I watched with alarm as over the next several hours jagged red triangles shot up the leg. By three in the morning, the angry redness had reached my crotch and I was as delirious as a college student with a spiked drink. My last coherent action was to bang on the door of director Rachel Maguire. The crew raced me to the Bali SOS clinic; upon arrival I passed out and remained unconscious for the next five hours. The "doctor" in Flores had done a shockingly poor job of cleaning the wound before sealing the flesh with the tight stitches, creating the perfect environment for the most lethal types of anaerobic bacteria. Untreated, this would have surely killed me. I reflected upon the irony of debunking Komodo dragons using bacteria as a weapon, then almost dying myself from infection.

Once I was up and hobbling around on crutches, we completed the filming by going to the superb Rimba Reptile Park. There we went into the enclosure of the giant captive-born-and-raised Komodo dragon named Monty. This leviathan really drove home to me just how intelligent these animals are. When I gazed into his eyes, he gazed back with brown eyes filled with sentience. He also, however, took an unsettling interest in my leg, obviously smelling the blood that lay beneath the bandages, so as soon as my portion of the sequence was complete, I wisely removed myself from the enclosure and chatted in the sunshine with my British friend Jon Griffin, who lived in Bali and ran a reptile export business with our mutual friend Duncan McRae.

Upon returning to Australia, I wrote up a Komodo dragon venom evolution paper that was accepted for publication by the esteemed scientific journal *Proceedings of the National Academy of Science.* My transcriptome analyses had shown that the venom was as complex as that of the Gila monster. In addition, my students worked in the laboratories of collaborators Paul Monagle and Wayne Hodgson to demonstrate the venom's effects of blood coagulation blockage and blood pressure lowering—effects entirely consistent with what I had observed in the field.

While there had been certainly no small amount of controversy when the first paper on the single, early evolution of reptile venom came out in *Nature,* nothing prepared me for the reaction to the Komodo dragon–specific study. This was best captured in an article by the esteemed science writer Carl Zimmer in the *New York Times* of May 18, 2009:

Chemicals in Dragon's Glands Stir Venom Debate

The Komodo dragon is already a terrifying beast. Measuring up to ten feet long, it is the world's largest lizard. It delivers a devastating bite with its long, serrated teeth, attacking prey as big as water buffaloes.

But in a provocative paper to be published this week, an international team of scientists argues that the Komodo dragon is even more impressive. They claim that the lizards use a potent venom to bring down their victims.

Other biologists have greeted the notion of giant venomous lizards with mixed reactions. Some think the scientists have made a compelling case, while others say the evidence is thin.

Biologists have long been intrigued by the success Komodo dragons have at killing big prey. They use an unusual strategy to hunt, lying in ambush and then suddenly delivering a single deep bite, often to the leg or the belly. Sometimes the victim immediately falls, and the lizards can finish it off.

But sometimes a bitten animal escapes. Biologists have noted that the lizard's victims may collapse later, becoming still and quiet, and even die. For decades, many scientists have speculated that the dragons infected their victims with deadly bacteria that lived in the bits of carrion stuck in their teeth.

Yet others have always been skeptical of the bacteria hypothesis. "Your average lion has a much dirtier bite," said Dr. Bryan Fry, a biologist at the University of Melbourne. "It's complete voodoo."

Dr. Fry suspected that Komodo dragons were using venom instead. In 2006, Dr. Fry and his colleagues published evidence suggesting that some lizards share the same venom genes as snakes. They concluded that venom evolved about two hundred million years ago in the common ancestors of the lizards and snakes. Studying an Australian lizard called the lace monitor, Dr. Fry found proteins in its mouth that were produced by those shared venom genes. When he tested the proteins, he found that some could cause a rapid drop in blood pressure and that others stop blood from clotting.

Komodo dragons, like the lace monitor, are closely

related to snakes, which suggested to Dr. Fry that the largest of all lizards might be venomous too. To test his hypothesis, Dr. Fry climbed into a Komodo dragon's cage at the Singapore Zoo. "I dangled a rat above it and got it really excited," he said. As mucus drooled out of the lizard's mouth, he used a test tube on a pole to collect it.

Dr. Fry did not find any venomlike proteins in the mucus, but later he realized mucus was the wrong place to look. Studying other lizards, he discovered that the proteins were coming from a separate set of glands in their mouths.

A medical disaster offered Dr. Fry a chance to take another look at the Komodo dragons. During a mysterious outbreak at the Singapore Zoo, most of the Komodo dragons died. Dr. Fry and his colleagues were given permission to dissect jaw tissue from a terminally ill lizard and preserve the heads of two dead lizards for later study.

The researchers found the second set of glands in the Komodo dragon heads, and inside they found venomlike proteins. Tests showed that one protein keeps blood from clotting. Another one relaxes blood vessel walls. "It drops the blood pressure like a stone," Dr. Fry said.

He argues that Komodo dragons depend on these venoms for their success. A Komodo skull is light, so it cannot generate a powerful bite or hold down a struggling victim, the way alligators do. "You'd expect them to be strong, but they're not," he said.

Instead, Dr. Fry argues, Komodo dragons slice open their victims, adding venom to the open wound. "If you keep it bleeding and lower its blood pressure, it's going to lose consciousness, and then you can tear its guts out at your leisure," he said.

Some experts praised the new study, which is published this week in the *Proceedings of the National Academy of Science*. "This paper clearly demonstrates that the Komodo dragon is a venomous predator," wrote Nicolas Vidal, an evolutionary biologist at the National Museum of Natural History in Paris, in an e-mail message.

But Kurt Schwenk, an evolutionary biologist at the University of Connecticut, is not so impressed.

Dr. Schwenk finds the new mouth glands intriguing, but he considers most of the evidence for venom in the study to be "meaningless, irrelevant, incorrect, or falsely misleading." Even if the lizards have venomlike proteins in their mouths, Dr. Schwenk argues, they may be using them for a different function.

Dr. Schwenk also doubts that venom is necessary to explain the effect of a Komodo dragon bite. "I guarantee that if you had a ten-foot lizard jump out of the bushes and rip your guts out, you'd be somewhat still and quiet for a bit," he said, "at least until you keeled over from shock and blood loss owing to the fact that your intestines were spread out on the ground in front of you."

During my undergraduate education, my Scientific Philosophy honors thesis examined how individuals respond to major paradigm shifts. My basic conclusion was that those who did not shift their view in light of new data had the same basic psychological mindset as religious adherents who felt challenged—they displayed defensive behavior, rejecting all new arguments without assessing them with due diligence. That certainly was the case here. The article quoted Kurt Schwenk, an evolutionary biologist at the University of Connecticut, who had no experience in venom research or with Komodo dragons. I found his stark statement that the evidence for venom was "meaningless, irrelevant, incorrect, or falsely misleading" as unprofessionally unscientific as it was deeply offensive. By rudely saying the evidence was false and misleading, he was effectively accusing me and my co-workers of scientific fraud. He displayed a fundamental lack of toxicological understanding in his statement that venom wasn't necessary to explain the effects of a Komodo dragon bite.

Our central point, which he clearly did not get (either willfully or through simple ignorance), was that the teeth were the primary weapons and the venom was there in a supporting role, to potentiate the blood loss and shock-inducing effects—something that was well supported by our high-quality, peer-reviewed data showing venom components with exactly these toxic effects. I felt it was telling that he could not come up with a criticism of the data itself and seemed instead to be simply making an argument from authority—despite not having any venom research expertise. I tried to be philosophical, but at the end of the day I was extremely offended.

The bizarreness of unscientific, emotional responses took on a surreal quality when, in collaboration with the University of California, Los Angeles bacteria specialists Ellie Goldstein, Diane Citron, and Kerin Tyrrell, we reinvestigated the mouths of Komodo dragons. These are the people who had done the seminal studies of bite-wound bacteria, ranging from bites in prison to bites from wild animals. If indeed Komodo dragons were cultivating bacteria, their mouths should have higher levels than ordinary carnivores, regardless of whether they were captive or wild. In other words, they should be selectively causing microbial blooming. The Los Angeles Zoo and Honolulu Zoo were particularly supportive of the research and let us take samples from their prize animals, for which we were extremely grateful. This is, of course, the ethical justification for keeping animals in captivity—for the intellectual and social benefit. The Los Angeles Zoo even gave us access to hatchling dragons only hours old that had not yet eaten or drunk anything: the most crucial sources of data.

However, other zoos were not so supportive (I will not name or shame them, despite *really* wanting to). Some ignored our request, while others flatly turned it down and, most staggering of all, some even went as far as ringing other zoos to try to convince them not to be a part of the study. When we submitted the paper, we had the most incredible battle with one of the reviewers, who could not make a coherent scientific case against our data or interpretations, but still stridently argued for its rejection. Once the paper was accepted, we assumed we were home free, but then the journal sat on the paper for an entire year. Only after we had sent many emails showing how annoyed we were was the paper released. I was gobsmacked. This was a subject for which we were providing the first solid evidence to debunk a cherished and long-held belief that had never been supported by evidence. Yet, unlike the issue of climate change, there were no economic interests being threatened. Collectively, this was as irrational a reaction to science as I had ever encountered. Between Schwenk's unprofessional response and the zoological institutions' illogical reactions, I was quite cynical by the time the dust settled. My academic conclusions in Scientific Philosophy as an undergraduate did not prepare me for the reality.

Not long after this, I was cleaning out the cages of the pair of Bell's phase lace monitors that were part of my collection of sixty giant lizards. My attention was distracted by the male when I absentmindedly made a small jerking action to flick away a fly, so I didn't see the female sprint across the cage. She latched on to my hand and jerked her head back, just like a Komodo dragon would with a deer leg. As the teeth cut my flesh to the bone, they made the ratcheting sound of a steak knife being dragged across a T-bone steak. Crimson ejaculations of arterial bleeding from my right index and middle fingers rhythmically painted the wall like a Jackson Pollock artwork. Having had many flesh wounds in my life, my first thought was, "Ah well, at least it's not a death adder bite. How bad can it really be?" It turned out I had grossly underestimated the gravity of the situation. In both fingers the tendons, nerve bundles, and arteries had been completely severed. All that was securing the two digits were the small ligaments on the side, where the finger bones met.

Upon arrival at the hospital, I had to explain how I came to be presenting at 7 a.m. on a cold, rainy, Queen's Birthday holiday in Melbourne with a hand destroyed by a large tropical lizard. I tried to convince the Indian locum to simply send me straight to surgery; that if I unwrapped the towels from my hand and unclenched my fist, I would unnecessarily start bleeding. He climbed up on his high horse and said in a thick Delhi accent, "I am the doctor here and you follow my instructions!" "Fine," I thought to myself. "If I'm going to bleed, you're going to suffer too." So I unwound the towels and unclenched my fist with my fingers spread, palm facing directly at him. Twin spurts of blood lasered out like Spider-man's web streams and splattered all over his shirt and face. He picked the wrong day not to be wearing a lab coat.

As predicted, off to surgery I went. It took a pair of plastic surgeons four hours of microsurgery to put my hand back together. In order to keep the tendons from pulling apart during the long healing process, my hand was in a brace and strapped to my chest for the next two months. During this time I had to do everything with my left hand, which drove me nuts. For example, learning how to wipe my bum with my left hand was a messy exercise in frustration. One-handed typing also took on a whole new meaning. I went through the swearing-in ceremony to receive my Australian citizenship with my arm still in the sling, my right hand resting above my heart. At least I looked patriotic during the anthem.

9
THE FROZEN SOUTH

My heart warmed at the email I received notifying me of the acceptance of my grant application to the Australian Antarctic Division. I was going to be able to lead my own team during the international collaborative Census of Antarctic Marine Life. Anything venomous that came up was ours. The Viking in me grew longer horns and the loincloth acquired quite a few more layers. I was going to the furthest ends of the world. While I had been far north up in Norway, nothing could match Antarctica. I had no idea what would be the dominant venomous life form down there, but wherever there is life, there is venom.

First, I had to complete the medical. "This is going to be amusing," I thought, when I saw that I had to undergo a psychological evaluation. The questions mostly had to do with how a person would cope with variable environments that were extremely different from the norm, and with social isolation. Since I was approved, they could only be selecting for antisocial, career-obsessed nutjobs!

Then it was off to Tasmania for pre-training and gear fitting for the team of Eivind Undheim, his wife Hanne Halkinrud Thoen, and myself. We packed up the wagon and drove down to the wharf in Melbourne and boarded the *Spirit of Tasmania* ferry. Once in Tasmania, the drive down to Hobart was interrupted periodically to check out road-killed snakes and Tasmanian devils. We deposited the gear at the depot to be loaded onboard the *Aurora Australis*, the 310-foot-long flagship vessel of the Australian Antarctic Division. Next came training. After covering the logistics and sampling schedule, the seminars concluded with the ship's doctor giving us a highly entertaining description of the simple physics of how quickly our bodies would lose body heat if we fell into the water, and thus how lethal it would be. Seawater freezes at a lower temperature than fresh water, so the water would be several degrees below that needed to freeze a tray of ice cubes. The room full of extremely energetic and enthusiastic biologists appreciated this macabre data.

Our voyage was slightly delayed when, during the final stress-testing of the engines, the starboard thruster underwent a spectacular self-destruction. The change in schedule meant two weeks less at sea, but we would still spend over two months without setting foot on land. The shortened schedule also meant a more restricted research program, with geo-coring the major loser, not because it is not terribly important—it is—but because it is slow and cumbersome to bore out geological samples under such inhospitable conditions at extreme depth. Largely dropping that research program would, however, allow for the full suite of biosampling to be conducted, and there would still be some scattered shallow-depth core samples taken over the course of the voyage. The reduction in crew members also meant that program leaders such as myself had cabins to themselves—a true luxury on any ship.

We polar-suited up and put out to sea. There were many friends and family members along the pier as the dream became, for all, a reality. As skies were clear and seas calm, we had at least four days of easy steaming ahead of us. Our first meal lived up to the legendary reputation of the chefs on this vessel. Fine dining while in the Southern Ocean seemed extravagant, but we would need all the nutrition for the physically demanding work ahead of us. During the trip down, we had ample time for the crew to get to know each other, and to explore this amazing ship. We discovered a vast empty space in one of the holds, and used that for soccer matches. While the ball was airborne, the ship would roll, thus giving the impression that the ball was moving in the most insane trajectory. We could bend it like Beckham but then it would bend back again with a corkscrew twist.

We got very excited at the first ice. Just like when losing one's virginity, looks didn't matter much for this first time. That it was there was enough. Such was our ice. Only a couple of feet across and all worn out, like a southern beauty queen turned meth-head. But our ice steadily grew in size and beauty. Icebergs of all colors cruised by—green, blue, red, hard white, and all variations in between. All a result of the varying mineral content of the water. Some were truly ethereal. When we finally crashed into proper pack ice, it was as close to a religious experience as I am ever going to have.

During the long haul to reach the pack ice, we had seen many birds, mostly petrels and albatrosses. But as we reached the ice, there was an explosion in the numbers and species. For the most part, I had no idea what they were. Despite my spectacular lack of knowledge about Antarctic wildlife in general, and the birds in particular, I was, of course, able to recognize penguins. They were absolutely adorable and hilariously clueless about what a boat was. Suited up in their tuxedos for a big night out, they would watch the boat with a total lack of comprehension until the giant icebreaker would collide with their icy home, launching them acrobatically but harmlessly into the ocean. The seals were a bit more switched on and would gracelessly flop into the water as we got near.

We used metal chains to lash a ten-foot-long plankton seawater tank onto the helicopter deck and rigged up two pumps: one carrying water heated to almost boiling by running coils that accumulated engine heat; the other carrying icy seawater. As we were all rostered twelve hours on and then twelve hours off, the hot tub was occupied twenty-four hours a day. It was a whole new experience—basking in the surrealism of having light snow committing suicide against steamy flesh, while serene whales swam benevolently alongside. Whenever we anchored, the whales would come abreast and stick their barnacle-encrusted heads out of the water to gaze at us while we gazed at them. The days steadily grew longer, until the sun never really set: it dipped just under the horizon and the waters turned to blood for a few hours. This was my favorite time of day, and I drew up my team's duty roster in such a way that I was on deck working the nets during this period.

The diversity and density of the Antarctic underwater life blew my mind. The tropics are always portrayed in documentaries as marine hotspots. They are, of course, but this viewpoint is skewed, as those regions are much more accessible and amenable to research than Antarctic waters. The sheer quantity and diversity of life in the Southern Ocean rivals anything I have seen on my expeditions to the Great Barrier Reef, Coral Sea, Asia, the Caribbean, or other locations in between.

Everyone met the first nets with much anticipation. Even those not rostered on deck were present to get the first glimpse of life under the ice. Benthic trawling was our primary method of collecting. The purpose of the trip was to survey the little-known life of the eastern Antarctic, and this was the most effective method. The nets were linked to a very heavy boom that allowed them to plummet to the bottom and stay pinned to it while being dragged along. This enabled us to efficiently collect all life present in the stretch of ocean floor. The first drag was mostly crinoids—the strange echinoderms—and also bryozoa, the coral-like life that is completely unrelated to coral. It was readily apparent that my usefulness in the sorting trays was going to be rather limited: I could assign most things to the kingdom Animalia only because they moved, and that was about it. So I set myself to working the nets and shoveling the collected material into large washing bins. Here mud was removed and the samples sent to the sorting tray conveyer belt, where specialists in different taxa divided them up by kind. The average net load was over one ton of material, so it was backbreaking, strenuous work. Exactly my style: a good workout while gazing upon the most breathtaking of sceneries.

The various species of ice-fish that came up in the nets consistently lacked any sort of reinforcement of the spines that stabilize the dorsal fins, indicating a total lack of venom. Similarly, the sea spiders appeared to be totally non-venomous, despite being another obvious inspiration for H. R. Giger in designing the face-huggers for the *Alien* movies. Pay dirt, however, was struck with the octopuses. The work I had been doing with octopus species from Australian temperate and tropical waters had revealed that venom was a shared characteristic of all octopuses, so any octopus counted as a venomous species. Not only were the octopuses present in abundance, but they also seemed extremely diverse. Knowing nothing about the taxonomy, we kept the venom glands from each specimen separate from those of any other and also took a DNA sample from each one, so that once back onshore we could determine their evolutionary relationships and link these to changes in venom. I was particularly interested in two types that looked virtually identical but differed sharply in their venom system. One had a massive beak and tiny glands, indicating that this type killed its prey using mechanical damage as the primary weapon. The other, however, was quite the opposite: it had a tiny beak but massive venom glands, indicating that its primary weapon was venom.

About three weeks into the trip, we ran into the unique type of massive polar storm that rightfully gives Antarctica the reputation of having the most extreme weather in the world. As the wind and waves steadily grew, even a ship as stately as ours started to feel the effects. Before long, the entire ship was shuddering and shaking from the impact of the huge waves. On the bridge of the boat, I was mesmerized watching these behemoths approach. At the peak of the storm, the waves were so large that they were crashing over the bow of the ship, which meant that they were at least sixty-five feet high. Every now and then a monster would hit us that was over one hundred feet. Suddenly our boat did not seem so big. Water would churn down the deck before chaotically pouring out of the side drains. This created a very specific hazard, so all hatches were sealed and on-deck activities suspended. Many people suffered from violent seasickness and the ship's doctor was extremely busy. I gorged myself on the extra rations that became available as many people were declining food, and largely spent my time in the sauna.

The storm cleared as quickly as it had appeared. The air was saturated with small ice crystals, resulting in a halo around the sun. We set up to do some extremely deep trawls now, with the nets reaching depths exceeding one and a half miles. It took over thirty minutes to drop the nets, but we spent four hours bringing them up! The super-deep drags mostly brought up otherworldly-looking flounder-shaped amphipods and tadpole-like fish.

Octopuses also showed up in the nets down to a depth of nearly one and a half miles. Moving back up the abyssal slope to about a quarter mile deep, we started getting more and more octopuses, until it finally happened: we had caught a giant octopus. I didn't even know Antarctica had its own, independently evolved species of giants. And a bonus—it was bright red. The giant, bell-shaped mantle was the size of a beach ball, and when I lifted the octopus up above my six-foot-three-inch self, the tentacles still reached the floor. The venom glands were flat milky jade discs nearly four inches across. In short order, two more giants came onboard in the nets. The particular microhabitat that we were surveying at this time was obviously the preferred one for these magnificent animals.

High on adrenaline, I was lost in my thoughts as I strode across the deck, "From Out of Nowhere" by Faith No More searing my eardrums. Not paying attention to where I was going, I walked at full speed into the two-inch-diameter metal cable that was under the tension of another trawl. I smashed face-first right into it, catching the full impact on my right cheekbone. It fractured with an all-too-familiar feeling. As we were still only halfway through the trip, and there was nothing to be done for this non-critical injury, I just kept it to myself and quietly dealt with the pain.

This wasn't the only injury I accumulated. The glass sponges and bristle worms were the bane of my existence as their fiberglass-like spines would penetrate even gloves and cause great physical irritation. I also slightly tore a plantar tendon in my left foot while doing yoga one day. I was deep in a lunge position stretch, leaning over my front left leg, when a large wave hit the ship sideways, causing me to lose balance, tipping forward and putting incredible strain on the tendons of the leg. I was also getting tendonitis in my forearms, but this was a simple mild irritant.

However, two days before the end of the trip, when lifting a mud-filled net, I tore the tendon that had been repaired in my right index finger following the destruction by the lace monitor a half year earlier. This was an incapacitating injury that could not be ignored. We had nearly finished the voyage, however, so I was able to just wrap it up and get through the final session, knowing full well that I needed to have surgery on it immediately upon my return to Melbourne. Using the satellite uplink, I communicated with a specialist surgeon I knew who treated a lot of Australian Rules football players and their incredibly screwed-up fingers.

At the completion of the incessant trawling we sank a net full of styrofoam cups to a depth of two and half miles. Everyone onboard had made cups of various designs, including very intricate ones prepared pre-trip by some. The incredible pressure at this depth caused the air in the cups to be compressed out, warping them to a small fraction of their previous size, with the irregular shaping adding a unique twist to each of the designs. I used the opportunity to make one for my parents. Now the tiny funny-shaped cup with ugly drawings is a souvenir in their house and I wonder if the guests think it was made by a kindergarten pupil as a Mother's Day gift.

The next day we set sail to return to Tasmania. The *Aurora Australis* docked in Hobart just before the completion of the iconic Sydney to Hobart Yacht Race. We were able to see the spaceship-sleek vessel *Wild Oats* come in with daylight between it and all other contenders. Back on solid land after two months at sea, with our equilibrium already shot, we took advantage of this loss of balance to cloak the effects of our liberal sampling of the wine festival that accompanied the yacht race.

With a bit of time to kill before heading back to Melbourne, despite the injury to my finger, I took the opportunity to dive at my favorite spot off Bruny Island to collect rock lobsters. This particular place is quite near a seal rookery and it was always a delight to play with these mischievous torpedoes. On this particular day they were a bit more skittish than usual, but I thought nothing of it since seals are pretty mental creatures and thus wildly unpredictable in their behavior. As I was working the rocky underwater slope that led into the deeper water, a school of two-and-a-half-foot-long mullet passed by, cavorting and shimmering like kamikaze pilots on their way to a happy ending. Less than a minute later, they came flying past me in a disorganized, chaotic stream. Watching them flash past, I turned back to face the deep green gloom just in time to see a six-foot-tall tail flick ten feet away from me, as a sixteen-foot, very bulky great white shark swam by, close enough that I felt the wake of its passing. Slowly easing back into the rocks, I did my best to still my frantically beating heart. Holding my breath so that there weren't any telltale bubbles, I slowly, stealthily worked my way into the shallows, taking my bag full of tasty rock lobsters with me. Walking slowly to the car after the excitement, I almost completed the day by stepping on a three-foot-long, patternless, black-colored tiger snake that was in the shade of the car.

After the eventfulness of the Antarctic trip, it was back to Melbourne, followed by surgery number two to fix the tendons sliced by the lace monitor lizard. This was followed by another frustrating recovery period, but at least this time my hand was not strapped to my chest. I just had to wear a metal brace for a few weeks. During this period, Eivind, for his Master's research under my supervision, worked diligently on the samples we collected in Antarctica, resulting in two great papers for the team. The first showed that the venoms of Antarctic octopuses have been evolutionarily selected to be most active at freezing temperatures, actually becoming less efficient at warmer temperatures. This points to potential problems for such animals in adapting to warming oceans. The second paper focused upon the genetics of the specimens, revealing four new species of Antarctic octopus, one of which was so divergent that it will be in a new genus. All in all, the most productive and satisfying field expedition of them all.

10
EL GRINGO

I returned to the United States, after many years away. In this post-9/11 world I noted a dramatic increase in the number of armed officials, some of whom wore uniforms of agencies I did not recognize. My luggage included various pieces of scientific equipment, some of which were very strange in shape, in particular the vapor shipper for storing and transporting venoms and tissues at negative 328 degrees. In its protective travel case, it looked like a mushroom, a Dalek or, to other eyes, a small thermonuclear device. Naturally, this could not fail to draw attention.

This was my thought as I was pulled up at passport control at LAX. I was escorted off to a side room where very stern people took digital copies of my fingerprints for searching against a database. I had no choice in this. The details from both my passports were also entered in. This was all done in complete silence while one agent in particular stood rather close to me. I was thinking this was a bit excessive for just a weird piece of gear, which had actually attracted scant notice. However, I noticed confusion creep across their faces as they wordlessly turned from the monitors and looked me up and down. Then one of them came over and asked me to roll my sleeves up. He said, "You don't have any tattoos there." I wondered why he was stating the bloody obvious.

He then asked me what I did for a living. I replied that I was a researcher in Australia on snake biology. He replied that the file said "biotoxins." I kept a poker face while internally my mind reeled. The paranoid little Norwegian gnome in my brain hissed, "See, I fucking told you, those damn rock trolls are out to get us! Never forget, even paranoids have enemies."

It confirmed something that I had long suspected: I featured prominently on security agencies' lists due to the nature of my profession. Any trawling of metadata will have in the explicit search file words such as "neurotoxin," along with many other words which feature prominently in my daily emails. Add to this my tendency to blithely wander through some pretty politically unsettled areas of the world and emerge (mostly) unscathed. The tracking of my Australian passport would reveal the most random travel pattern to some of the most non-touristy spots in the world. I used my United States passport, however, only for entry/exit to and from the US. I had figured that with all of this combined, I would be caught in the data filtering by intelligence agencies. It was only natural, since I could kill with any number of toxins or even a combination of unrelated chemicals. If a particular effect was desired, I also knew which ones were detectable and which ones were not.

So I was not surprised at this confirmation that one or more of the countless international security agencies had picked me up on their radar, whether it was Interpol, domestic intelligence in England (after the airport fiascos and also milking a cobra live on television), or ASIO, the Australian intelligence service, since I was at this point consulting for different government committees about venoms, ranging from biodiscovery intellectual property protection, to environmental listings, through to the use of venoms as bioweapons.

But it turned out to be something much more mundane: simple identity theft. I had had my wallet stolen during my undergraduate years. The person who stole it tried to use my Oregon driver's licence when pulled over after drunkenly hitting several cars in Medford. This, of course, was a fail, since he was a shortish Hispanic guy while I am six foot, three inches, and as Aryan as they come. The police station had returned my wallet and licence (the money was, of course, nowhere to be seen). Boy Wonder didn't show for his court appearances, so a warrant was issued for him under his own name and also under his aliases, which were all spelling variants of my name, including my Norwegian middle name, Grieg. This was completely ridiculous, since the same people who entered the aliases had been the ones who returned my wallet to me. It meant that my name was now flagged in all law enforcement agencies that had access to this system or derivatives thereof—a completely ridiculous state of affairs. I was able to straighten it out in the airport security computer, so that I would not be held up again for this reason. But there was no guarantee that some state cop somewhere wouldn't have it come up if I were ever pulled over while innocently speeding.

At least this incident confirmed my hunch about being noticed by intelligence agencies as a result of my toxin expertise. It didn't really bother me; I had nothing to hide. And besides, it meant that if the world got invaded by an alien race of venomous animals, I would be the one they would call upon to heroically save the tattered remnants of humanity from the scourge. Naturally, I would end up with the hottest female among the ten thousand people left alive at the end of the war. Michael Bay can direct the movie. Like a cockroach, he would survive even a direct nuking from space.

My purpose for being in Los Angeles was to pick up a car and drive north to San Francisco to lead a seminar at the California Academy of Sciences, where Joe Slowinski was working when he died. It was strange to be giving a talk there. As a devout atheist I, of course, have no belief in any sort of afterlife. Regardless, it was like the ghost of Joe was there in the questions not asked or the comments not given. An existence whose genesis was a vacuum. My talk was very well received and the questions lucid and insightful. It was very nice to be giving a talk at a place I had visited during my formative years. This and the Steinhart Aquarium were favorites of mine during the five years we lived at Hamilton Air Force Base, north of San Francisco. It would have been even nicer if my mate Joe had been there.

Joe was now a public figure. In the years after his death there had been much reporting and analysis, culminating in the book *The Snake Charmer* by Jamie James. While he has a name that would suit the latest soulless clone in a UK boy band, he is in fact a very solid reporter who had a reputable career before wisely deciding he could write books while living the good life in Bali. He interviewed quite a number of people across a broad spectrum, and thus was able to do justice to an emotionally charged topic. I was able to relate to him how I felt during my death adder envenomations and the euphoric sensation that had resulted. Krait venoms act much the same way but are even more potent, since they bind at the point where a nerve waits for a message from another nerve in a series, so that the message never gets through. They also have similar toxins to a death adder, which block the receptor that sits on a muscle waiting for the message from the nerve. So with a krait bite, you are double-teamed. Usually you have to pay extra for that kind of action. And you do pay—your nerves are getting hammered twice as hard. I made it clear to Jamie that I could only speculate, but based on my experience I reckon Joe would have been feeling the same sort of delicious sensory distortion that I had. Perhaps even more intense. If this was the case, it was not a bad way to go.

After my talk, Jens Vindum gave me a tour of the preserved specimen collection. I wasn't expecting the emotional lightning bolt that struck me when I came across the exact krait that had killed Joe. It had been preserved and brought back when the expedition abruptly ended after Joe died. It was very interesting to talk to Jens about the lingering legacy of Joe's death. Whatever the level of personal grief, his death had profound impacts upon the ability of researchers to do field research of any sort, let alone in remote locations with potentially lethal animals. This was most acutely felt, of course, at the California Academy of Sciences, where there was the inevitable overreaction, resulting in layers of stifling bureaucracy.

From there I was off to milk beaded lizards with my mate Howard "Howie" McKinney. Howie is a clinical pharmacologist with a legendary reputation in the scientific community that investigates hallucinogenic drugs. His authority came from extensive personal experience. People love having him around for the weirdness that gravitates towards him like a moth to a flame. So we added to the lore by going up to the psychiatrists and getting them to give us a few of the special question-mark-shaped soft plastic devices they had specifically for psychotic people having a fit, who can bite down on such a device as hard as they want without doing themselves harm. We were going to use them to milk beaded lizards that were in the collection of extremely talented private keeper Steve Angeli.

These bits of rubber would deform as the lizard bit down. I had noted that these lizards had a much more powerful bite than a monitor lizard. When they clamped down on a heavy leather glove, the teeth would grip but not penetrate—they were too narrow and deeply grooved. They were more like syringes than the serrated knives of the giant lizards. But the giant lizards have very weak bites because they need light heads that they can swing quickly while pursuing fast-moving prey. The beaded lizard, and its smaller sibling the Gila monster, both have broad heads with very large jaw muscles. Monitor lizards are lithe and bird-like. But the beadeds and Gilas feed on rodents underground or on the nestlings of ground-breeding birds. So they occupy a niche of close combat, in which brute force is more useful than agility. This means they can hold their prey and chew the venom into the flesh, while compressing the flesh rhythmically to move the shock-inducing venom along. They are heavily armored, with bones in each scale, and thus able to resist the bites of even enraged female rodents, which they would eat first before devouring the jellybean-like babies.

On the drive back, Howie recounted to me a story about an event that was easily the most extraordinary story involving a venomous animal I had ever heard. In the late 1980s on a quiet weekday evening in the emergency department of a San Francisco hospital, a phone rang in the Poison Center:

Howie: Hello, can I help you?
Caller: Yeah, I, umm, got bit by my snake.
Howie: Okay, are you all right?
Caller: Yeah, I'm basically okay.
Howie: Okay, what kind of snake, do you know?
Caller: Oh yeah, it's a rattlesnake.
Howie: *Crotalus*, right?

Caller: Yeah, a northern Pacific rattlesnake, you know, one of the locals.

Howie: Umm, you sound very calm. Are you okay?

Caller: Oh yeah, I'm just fine, but I think I maybe got some venom this time.

Howie: This time?

Caller: Right, uh, I get bit all the time.

Howie: I see. Where were you bit?

Caller: Right here in my home.

Howie: Okay. Actually, I meant, where on your body were you bit?

Caller: Oh, sorry, uh, you know, right next to my scrotum.

Howie: Uhh, okay, I see. How long ago?

Caller: Maybe about half an hour.

Howie: And you said you thought there was some venom. What makes you think that?

Caller: Well, you know, it stings and hurts at the fang mark and I have that minty taste in my mouth.

Howie: I see. I can call an ambulance to get you to the hospital.

Caller: That's okay, my friend is driving me, we're about to leave. Actually we are only about ten minutes away from you.

Howie: Then we will be expecting you shortly. Sure you don't want an ambulance?

Caller: No, I'll be okay.

And he gently hung up his phone.

Certainly one of the more unusual phone calls, even for a place that receives unusual calls all the time. Most snakebite calls involve a frantic caller, but the extraordinary calmness here was noteworthy. Plus, the caller recognized the genus name and quickly and correctly identified the species. This was an experienced "herp" person, he did not sound intoxicated, and despite several very uncharacteristic elements in his story, he was so calm and matter-of-fact. A veteran of ER departments, Howie had long ago learned to take these scenarios as they present themselves, seek the internal logic in the situation, and attempt to evaluate the patient and plan responses to help them.

Howie told the story to the triage nurse and physician on duty, and while he obtained antivenom supplies, the staff prepared a room for the caller's arrival. All of the staff had seen snakebites before, and constantly witnessed bizarre behaviors that land people in emergency departments, but this one had engaged the interest of the entire staff. The police, several firefighter-medics, and doctors all were curious to see this patient.

And a few minutes later two slender men in their 30s, well-groomed, wearing jeans and T-shirts, walked into the emergency department, one man assisting the other walking with a slight limp, and the limping man holding a large capped glass jar containing, yes, one neatly coiled rattlesnake.

The patient was swiftly placed sitting up on a gurney, and as the gurney was wheeled back to the treatment area, vital signs were taken, clothes removed, and hospital gown draped around him.

The snake was indeed identified as a northern Pacific rattlesnake, very much alive and calm, coiled in the bottom of the jar. Snakebites occurring outdoors in appropriate environments and circumstances for encountering a snake are nearly always an indigenous species. In those circumstances local emergency departments typically had supplies of appropriate antivenom for the local species. But "pet snakes" could be any species including "exotics," which can be very venomous animals for which antivenom may be locally unavailable, requiring the hospital to sometimes obtain supplies from foreign countries. Networks for clinicians to obtain antivenoms and expert consultation that are quickly available nowadays simply did not exist in the 1980s. Quick positive identification of the offending snake is still of great benefit in planning treatments for the envenomed patient.

This patient, now calmly resting on his gurney in the treatment room with the rattlesnake in the jar on a bedside table, was being subjected to rapid-fire volleys of questions while intravenous lines were started and blood for lab analysis was collected in various tubes. Do you have pain here? Any nausea? The staff searched for any evidence of ecchymosis near the single fang mark, swelling in the area of the bite, and other physical examinations to document an envenomation requiring antivenom administration. For a "dry bite" wherein the snake strikes, leaves fang marks, but does not expel any venom, local wound care is usually sufficient without the need for any antivenom.

Having determined that this patient did indeed have a moderate envenoming, they commenced the Wyeth antivenom preparation. This antivenom was notoriously difficult to reconstitute, the lyophilized ("freeze-dried") powdered mixture required about twenty minutes of swirling in the vials to get the powder into solution so it can be injected by intravenous infusion. While an initial dose of six vials were being swirled by staff in the room, the patient maintained a sharp interest in the activities, even requesting to swirl two vials of his dose into solution himself. Upon their first encounter in the hallway, the patient and Howie engaged easily in conversation, and throughout all the activities he stayed focused on Howie. With his initial evaluation complete and antivenom preparation in

progress (yes, in the 1980s emergency department drug preparation was frequently accomplished at the bedside, a circumstance
that would be very unusual today), Howie began asking him about
the circumstances of his bite.

Quite a large "audience" of people had gathered just outside
the doorway to his room, and all fell silent, listening, as the patient
and Howie continued their conversation.

> **Howie:** So, you seem to know more about snakes than our
> typical patient.
>
> **Patient:** Well, snakes have kind of been, you know, my
> hobby, since I was in my teens. I just enjoy watching
> them and working with them.
>
> **Howie:** So, do you mind if I ask what happened this eve-
> ning that resulted in your being envenomed? Were you
> feeding the snake?
>
> **Patient:** Uh, no.
>
> **Howie:** I see. Maybe you were handling the snake?
>
> **Patient:** Uh, yeah, kind of handling it.
>
> **Howie:** Maybe milking the venom?
>
> **Patient:** Oh no, no, I had done that earlier in the week.
>
> **Howie:** Uh huh, do you collect the venom for donation to
> science? Like give it to a research scientist?
>
> **Patient:** No, I don't. Actually that's a great idea, just never
> thought of it. No, I milk them just to be safe.
>
> **Howie:** Safe?
>
> **Patient:** Right. Uh, you know, so there's no venom when I
> handle them.
>
> **Howie:** So, you free handle your rattlesnakes?
>
> **Patient:** Oh yeah, all the time.
>
> **Howie:** Uh huh, I see. And your scrotum, that's not the

usual bite site. Actually not unheard of, but, you know, unusual. Was the snake maybe in your lap?

Patient: Oh no, nothing like that. He was on the table in front of me.

Howie: Okay, on a table in front of you. Like, just lying there loose?

Patient: Well, really, he was all coiled up ready to strike.

Howie: Really? And you, uh, were you wearing any protective clothing? Using a hook?

Patient: Oh no, nothing like that. Actually, uh, well, uh, actually I was naked.

At this point the entire audience was, very uncharacteristically for an emergency department, utterly and totally silent. Not a single word could be heard, except for Howie's conversation with this patient. He maintained his reserved, unemotional deadpan intonation, a very matter-of-fact delivery. Throughout their encounter, it seemed he had developed a trust in Howie, kind of a bond. He seemed reluctant to speak with other staff, but continued his narrative with Howie's gently prompting.

The story that emerged was, well, let's just identify him as an entertainer, of sorts. To the growing disgust of the audience, he explained that he charged admission to people to watch a sex show. He performed naked with the rattlesnake positioned on a table at "scrotum height" about two feet in front of him. As he performed various activities, the show was brought to a rousing climax as he masturbated while the snake would strike him around his groin.

Hence, naturally, the need to milk the snakes so there was no venom. Of course. Which also explained the "I think I maybe got some venom *this time*" comment. It was all adding up quite nicely. Bizarre, but it all fit.

The audience in the emergency department were by this time making loud comments, which included commentary about how utterly sick and depraved this patient was, how the neighborhood was not safe with him around, how society would be better off without such sick people, and various other comments along those lines.

As the emotional crescendo elevated around his room, the antivenom almost ready for infusion, the commentary was causing this patient considerable distress. Howie was still in conversation with him and Howie himself was becoming upset at his rude colleagues. So, as the taunts became more graphic and had developed a dynamic momentum of their own, Howie reached boiling point. He turned to the doorway, stopped swirling the antivenom, and shouted at the audience.

"You all need to just shut up *now*. Yes, this patient has an unusual occupation. And you are insulting and scaring him. And you're talking about how scary *he* is with his sick little sex show? You know what? In a way, he's just a guy who has found a way to make a buck, to get by, to survive. You know who scares *me* in this story? Not this patient, it is the weird twisted sickos in his audience! Those are the people who scare me!"

As they got back to the business of antivenom infusion and monitoring the patient, the friend took the rattlesnake in a jar home, and the scene became history. He was not even admitted to the hospital, just spent several hours in the emergency department being observed for any problems.

Later, several of the team and Howie were reviewing the extraordinary scenario they had just witnessed. Upon reflection, they agreed, that those who would *pay* to watch such a performance were indeed the sick ones.

Weirdness, after all, lives in the mind of the beholder. And perspective changes everything. But one can certainly have too much fucking perspective!

Once the seminars were all given and scientific tasks successfully undertaken, it was time to fly out. As customs rubber stamped my US passport I wondered to myself why none of them ever asked where I went to, since there were no other stamps. My Australian passport was packed with stamps and almost completely full, despite it having four times the number of available pages as my US passport. It was as though Australia expected its citizens to travel while, conversely, the United States did not expect its citizens to leave the country. This was, of course, the country that "elected" Bush, who did not obtain his first passport and travel out of the country until after becoming president. There should have been a rule that if he couldn't pronounce it correctly and find it on a map, he couldn't invade it.

Landing in Mexico City, I checked into the Four Seasons Hotel and headed down to the bar, where I had a decent beer, but the guacamole was so awful it might as well have been made by an Australian. Meanwhile, the piano man serenaded the largely empty lounge with a weird Spanish rendition of the Wham! song "Careless Whisper." The only guilty feet were those of the musician working the pedals as he managed to transform an already-awful song into something even worse, or *muy mal,* as the locals would say. After a peaceful sleep to the soporific sounds of gunfire, I headed to the Cuernavaca UNAM laboratory of my amigo Alejandro Alagón. The goal of this leg of the trip was to collect additional species for the lizard venom research, while also testing the waters of a new research area: vampire bats.

While Mexico was well known as having the venomous beaded lizard, our research had determined that there were many more venomous lizards than previously recognized, and this meant that there were a number of unique species available in Mexico to be studied. Interestingly, all of these potentially venomous species were locally referred to, along with the beaded lizards, as *escorpións*. It was yet another example of how indigenous populations typically know what's going on long before foreign scientists.

The two main targets were the arboreal species of alligator lizard and the crack-dwelling knob-scaled lizard, both of which were to be found in the cloud forests a long drive from Cuernavaca. This drive was among the most hair-raising of my life. For some reason, there were large arrows indicating when to drive on the normal side of the road for the country and when to switch to drive on the left, as in Australia. Visibility was severely compromised by the thick clouds that gave the forests their name, leading to constant anxiety about a head-on collision being imminent. I can only assume this was done to help the trucks navigate the steep roads more safely, but it was still terrifying.

With Tito & Tarantula's song "Strange Face of Love" playing on the car stereo, we reached our destination in one piece and went hunting for lizards with our guide Roberto Mora. We hunted for the knob-scaled lizards at night, when they would be in their refuges in the rocks, safely tucked away in the deep cracks that crawled across the boulders like the sun-damaged wrinkles on a beach bum's face. We would use a pencil light to shine in just enough light to see a lizard, but not so much as to startle it. Then, using a long piece of wire, we would tickle the very tip of its tail, as if a snake were sniffing at it from behind in the crack. This would invariably send it sprinting forward and arcing out of the crack into the air. We'd flail like a baseball outfielder trying to catch it. Despite our success rate being only about 10 percent, after a couple of nights' hunting we had the half-dozen specimens we needed for the research. These little lizards were fascinating—they were genetically intermediate between a beaded lizard and an alligator lizard, and looked the part: while their heads were those of an alligator lizard, their scales and their tails were those of a beaded lizard.

Next up were the arboreal alligator lizards, which proved to be frustratingly hard to find. The weather had turned and cool rain saturated the forest—hardly ideal conditions for reptile hunting. After three fruitless days, we were walking down a dirt road one morning when we spotted a tiny object far off in the distance. It was a person running toward us, but still over two miles away. Once he got closer, I could see that he carried two clear plastic bottles, one in each hand. When he reached us, I spotted flashes of green in each. Two Germans had been through in recent months and had created an artificial (and illegal) cargo cult economy for arboreal alligator lizards, which are prized on the European pet market. These lizards are now very rare in the wild, due to habitat destruction, so it is illegal to collect them without a permit, such as the scientific authorization we possessed. But the Germans had purchased quite a few before flying off with their illicit wares concealed in their luggage. They were stopped by Mexican border patrol on the way out, however, and arrested for wildlife smuggling. There is a cynical joke regarding this behavior: "Question: What's the first thing that happens when a new species of reptile is described? Answer: Two Germans buy a plane ticket." We were reluctant to encourage the economic dependency of remote populations like this one, but desperate for specimens for the research. So after some rapid-fire haggling in Spanish, we settled on a mutually acceptable price and went on our way.

The sun broke through the clouds, so we kept looking as the area rapidly warmed up. While we did not find any lizards, we did find something else I had long been searching for: the black-tailed horned pit viper. These snakes are typically found on the ground, but within an hour we found two, both basking at the top of three-foot-tall bushes, looking like little smoky brown clumps of dead leaves. Milking this species meant that I now had a venom sample from every genus of viper in the world, thus completing the set and launching a series of projects. I was extremely pleased.

The rest of my time was spent collecting scorpions. The part of Mexico we were in has an annual incidence of five hundred thousand scorpion stings during the four-month tropical monsoon season. One hundred and fifty thousand of these require life-saving, pain-relieving antivenom. The scorpions were present in plague proportions, so I'd zip my shoes and all other gear into packing cells whenever they were not in use. Collecting scorpions was extremely easy—you simply had to shine an ultraviolet flashlight around at night. The scorpion exoskeleton would reflect back like a nightclub ornament. As a safety precaution, we wore special protective glasses to prevent the strong UV light from damaging our eyes, which could lead to an increase in cataract formation.

Back in Cuernavaca, we collected vampire bats. The method of capture was to string extremely fine nets called "mist nets" across cave mouths during the late afternoon and then settle down and wait for nightfall. The caves were shared by a wide variety of bat species, most of which were the insect-eating microbats. These little guided missiles were fast, but not too bright. Upon encountering the nets, they would thrash violently and instantly get entangled. Vampire bats, however, are very smart animals, since they have to calculate an approach to much larger, potentially dangerous warm-blooded prey. They first fly near to the prey and land on the ground. The rest of the approach is through a series of stealthy hops and stilt-like walking motions. Upon encountering a mist net for the first time, they would freeze, swivel their heads around, immediately figure out the situation and actually walk down like a spider from Mars until they reached the bottom, drop off, and then fly away. So if we weren't fast, we would miss out on the capture.

It was complicated by the fact that the vampire bats were also extremely light-shy, so if we had our headlamps on, they would not fly out of the cave. This meant we had to flick our headlamps on for a fraction of a second every few minutes to see if there was a vampire bat on the net. This made the hunting that much more fun, as I love nothing more than a good challenge. It took us two nights to get the trio of vampire bats we needed for the research, while we plucked fifty microbats from the nets each night.

From there, I flew to Colombia, a country with a reputation for being extremely dangerous but with reptiles found nowhere else. This is because the narrow isthmus with Panama created a genetic bottleneck as snakes pushed south after invading North America over a land bridge from Asia many millions of years ago. Upon reaching Colombia, the animals were then filtered through a series of very large mountain ranges, producing a high number of endemic species. I landed in Bogotá and stayed the night ahead of catching a connecting flight to Santa Marta the following morning. While I was waiting for my flight out of Bogotá, my gear and I were thoroughly searched inside and out by airport security officers suspicious of my two large aluminum suitcases, the type favored by the CIA. They opened them up, scattered my belongings haphazardly and even knocked on the side panels. What they thought they would find was beyond me, but I was totally unconcerned since I knew there was nothing to be found. Amusingly enough, they ignored my snake-catching equipment and field guides. With an air of resignation, they allowed me to depart.

I then met up with Juan Manuel Renjifo, a herpetologist so cool he even named a really weird coral snake after his daughter. *Micrurus camilae* is unusual in that instead of the red and yellow colors being in separate bands, with black rings in between, the dorsal is colored orange/red and the ventral yellow in each of the two-toned color rings that alternate with a black ring.

Hunting for coral snakes is coincidently also best done in the same way as hunting something else I was after: armed spiders, also known as wandering spiders. These are the largest of the "advanced spiders," a type of spider characterized by small chelicerae, the fang-like structures from which spiders deliver their venom. These spiders are of the sub-order Araneomorphae, and are thus often referred to as araneomorph spiders. All are characterized by having small fangs that move inward like pincers. Primitive spiders (sub-order Mygalomorphae), on the other hand, have fangs that swing out vertically like a rattlesnake's fangs, traveling parallel to each other. Despite being large and fearsome-looking, mygalomorph spiders such as tarantulas are almost harmless to humans, producing typically no more than bee sting–like pain and swelling. The exception is the male of the Australian funnel-web spider, which has a bite that can kill humans. All other spider species lethal to humans are araneomorph spiders.

The wandering spiders are the biggest of all araneomorphs, with two-inch-long violin-shaped bodies, and legs as long as the body. They are powerful, agile hunters that will attempt to kill any animal, vertebrate or invertebrate, up to twice their size; they will readily attack tarantulas much larger than themselves. These spiders hunt by sight, not using a web. They can run fast, but also leap up to one and a half feet into the air, traveling on an arc that carries them forward by one and a half feet. People are bitten when they try to push them off the porch with a broom, and the spider runs up the handle and bites a finger. As many people receive antivenom for this spider as for any particular species of snake.

To catch both the wandering spider and the coral snakes I was after, Juan and I set about clearing large brush piles of oil palm leaves found on a research station that had its own plantation. On the commercial plantations, the amount of pesticide used kills off the entire food web because of the depletion of insects, leading to vast chemical wastelands of green sterility. But these particular mounds had been untouched for months and were now home to a vast array of creatures, ranging from microscopic life in the undergrowth, through to coral snakes and wandering spiders, which are apex predators in this microhabitat.

Clearing a mound started with first cutting away all the weeds, grass, and growth around the pile to create a three-foot-wide ring. Even the morning air of the Colombian wet season is hot and very humid. As the mound was cleared from the edges, whatever was in the mound retreated to the center rather than coming out into the open sunlight. Each frond had to be vigorously shaken and then closely inspected, particularly as some extremely cryptically colored and patterned insects can be almost impossible to see unless they move. The deadly lancehead vipers also are very difficult to see; venomous landmines is how I thought of them. As the pile became smaller, things got more exciting, until a certain critical mass was reached and myriad life showed up. Coral snakes were few and far between, but invertebrates such as wandering spiders, scorpions, and centipedes were abundant.

Colombia is a place of contradictions. It's one of the most visually stunning places in the world, with the kind of beauty that makes you weep just for the privilege of having cast an eye upon it. Beauty that only a master oil painter could capture, as photographs do not do it justice. However, paradise is being destroyed through malignant mismanagement. The lushness that makes it such a botanical wonderland also facilitates the growing of the two most evil crops on earth: cocaine and oil palms. It might seem strange to include these two in the same category. Before going to Colombia, I certainly would have thought they'd be opposites, with palm oil being the green-fuel "golden child."

In addition to sustainability considerations such as depleted soil or the environmental catastrophe that results from chopping down primary rainforests and replacing them with chemically saturated plantations, the social costs have to be considered too. Palms are replacing bananas on the existing plantations that were part of the huge US-agriculture business empire that flourished in Colombia during the first half of the 1900s.

Bananas not only provide steady work but they provide the workers with virtually unlimited, free high nutrition to take home to their families, to be used in a wide variety of recipes. In contrast, palms are harvested once every three to four years, so the work crews are rotated around the country by the companies. Thus, in the switch to palm oil the local communities have been economically devastated, while simultaneously unemployment has soared. Not only are the locals out of work, but also they now have to buy overpriced, low-nutrition, highly processed food. This has worsened the social disenfranchisement that was the impetus for the formation of the guerrilla movement, which had a noble social goal at the beginning. However, this noble movement became horribly corrupted along the way and now exists purely for the economic riches brought by drugs and kidnappings. The writing is on the wall for the worsening of the social situation of this magnificent country, which is very sad, as it has so much potential.

This social inequality is what makes Colombia so dangerous, particularly for foreigners. My field collaborators and I had just finished hunting for multicolored giant coral snakes in the La Victoria coffee plantation, located on the mountain range up from Minca, and were driving back down toward Minca. Suddenly, a dozen motorcycles with armed riders were circling the car while we paused at a stop sign. Being boxed in gave me a sinking feeling in the stomach. Juan spoke in excited Spanish, far too fast for me to follow, and held up snake bags. This seemed to satisfy—or at least confuse—our potential captors enough for us to be on our way.

The danger was clear and present. Only a few months before, the owners of the coffee plantation had similarly armed people invade, line up their staff, ask questions to determine identity, and then shoot dead one of the workers without warning. They were convinced that he was talking to an opposing group a few mountain ranges away, since they had been tracking all mobile calls in and out of their own area. In reality, he had simply been ringing his girlfriend, who happened to live in that area. I was very happy to leave Colombia and head to the relative safety of Brazil. While Colombia was stunning and the animals amazing, it was a country I vowed I would never return to.

My intentions for Brazil were to simply enjoy the land. No research quarry. Just going troppo in the tropics. Landing in Manaus after a series of flights, I could tell instantly that Brazil had a very different vibe to it. More cool, calm, and collected. Perhaps this was due to the sedative effect of the *caipirinha,* a potent, sugary lime drink infused with the local rocket-fuel liquor called cachaça. I was awesomely devastated by several glasses of this when meeting up with Dick Vogt, ahead of joining his research group upriver. Dick is a brilliant American herpetologist who has been leading a delightfully decadent existence since arriving in the Amazon decades ago and deciding to never leave.

We navigated the chaotic dockland and boarded a three-story people-mover boat to head up the Amazon. I was rapt as the first pink dolphins appeared—mermaids of the tannin-stained waters—as well as the caimans and iguanas, which created scaly ridges on the riverbanks. But the boat ride quickly lost its novelty and appeal; it was sensational for the first few hours, but by iguana number two million they became less mesmerizing. The reptiles might as well have been moss-covered rocks for the level of interest they stimulated in me. If someone had said that being on a boat on the Amazon would be boring, I would have looked at him or her as if they were as mad as everyone considered me to be. But imagine watching the same twenty seconds on loop for days on end. There was nothing to do but drain tiny can after tiny can of the truly awful Brazilian beer called Brahma Chopp, which I renamed Brahma Crap. Predictably, this led to me getting burnt like a British tourist on Tenerife. I went the "full lobster" with this one. Never go full lobster.

Three days later, after boat rides in successively smaller vessels, we were in small aluminum boats with little hamster-powered outboard engines, the sort that no self-respecting saltwater crocodile in Australia would pass up the opportunity to crush like a beer can. A comforting thought as I gazed at the waterline so close to the gunwales that a well-timed epileptic fit would be all it would take to capsize us. We had seen many black caiman by this point, the local crocodilian filling the niche scientifically known as "bad-assed motherfucker." As big as the benign American alligator that inhabits Florida golf courses, but much more vicious, the black caiman regularly made a meal out of village children.

The research site was a seven-hour hike once we left the last boat. Howler monkeys screamed like dementia patients while intricate lace-winged butterflies fluttered by. Or for me, just crashing like Mowgli through the water, since my ability to stand leeches was much better than my balance at any given time, let alone with a heavy backpack. The others found this amusing, but my logic was that if I went on to a slippery log bridge, I was almost certainly going to take a clumsy fall eventually. So if I was going to get wet anyway, it might as well be bipedally.

During the long boat ride up, we had slept entirely in the open air in hammocks. While we were moving, mosquitoes were not an issue, but now that we were camping in the Amazon forest, mosquito-borne disease was a constant concern. Before leaving Melbourne, I had made sure I was vaccinated against everything I could be, ranging from yellow fever to some pretty rare types of encephalitis. We were remote enough that the streams were drinkable—a luxury indeed.

Camping was as simple as stringing hammocks between trees and then cocooning them with fine-mesh nets to keep out the mosquitoes. The first two hours of darkness were the danger zone for the malaria mosquito, a quite delicate little creature that haunts the early evening. So about an hour before nightfall, everyone would curl up in their hammocks, safe behind the fine mesh netting cocooning them like messy spider webs. Everyone would bring a headlamp, a book, and anything to help them pass the time—various chemicals of choice and a fine whisky were passed around over successive nights. Whatever you had that night was all you were going to have for the next three to four hours. The malaria risk was extreme where we were, but we knew the flight patterns of this mosquito, which meant staying put during the defined activity period. I discovered that peeing into a bottle while in a hammock is a very challenging task that requires much concentration. The last thing I wanted to do was flip the hammock, go airborne, tear through the netting, and end up a urine-covered heap below . . . with malaria. Not my idea of a good time!

There were mosquitoes that spread other diseases, but where we were they were in much lower concentration than the malaria mosquito. So with normal precautions we were fine for other biting insects. The small black flies, with their oversized wings, were one of the most annoying animals I've ever encountered. They were only around good-quality fresh water. Bites invariably result in a violent reaction, ranging from small hives through to grotesquely swollen ankles. They prefer white meat. The local Brazilians would still get bitten, the wound being obvious, but the bites wouldn't swell or itch. This is because they had been bitten regularly since they were young and had developed antibodies to the anticoagulant proteins the fly spat into the wound to keep the blood flowing.

We were staying with a local indigenous tribe, who were cooking for us and helping to run the camp. One night, they shot a capybara relative and it tasted exactly how I imagined a large forest rodent would taste. Rat-kebab is not on my shortlist of future meals to cook for a date. We washed it down with lots of caipirinhas. Pickle the tastebuds with this liquor made from sugarcane in the tropics, and anything can be eaten!

I took full advantage of the space for solitude. While I was in Colombia, my male dingo Norton had been diagnosed with malignant lymphatic cancer, so I'd had to give the okay for euthanasia over a long-distance phone call. To do this without being able to say goodbye was terrible. It affected me deeply and I needed nothing more than some long walks through the jungle with only my thoughts for company. However, I discovered that one couldn't get more decompressed than tripping on some potent herb with members of an indigenous tribe. Their ritual, not mine—but that didn't mean that I couldn't enjoy it if invited. It would be rude not to!

I drank deeply from a bowl that contained a liquid which tasted like green smoke. But in a good way. Whatever this stuff was, it far exceeded anything I had ever had in the Netherlands. It even beat the joint recipe I used all my biochemical training to perfect in Amsterdam over several successive northern hemisphere summers: the final ground-up mixture was 50 percent dried marijuana with the rest made up of equal parts of dried psychedelic mushrooms, datura seeds, catnip, and blue lotus flowers. This drug neapolitan produced extreme visual hallucinations with the occasional bout of fractal vision, but none of the emotional intensity triggered by eating psychedelic mushrooms. The blue lotus flowers didn't contribute to the dissociative effects, but rather made the smoke as smooth as polished diamonds. The catnip gave it a delightful twist. If this is what the average housecat goes through, they are doing it right. But nothing in my chemically consumptive history compared to this Amazonian delight that was meant to induce a vision quest—which is just a fancy way of saying, "You are going to have a very long hallucination, white boy." The closest I had come previously was being off my head on Hawaiian woodrose seeds while playing with my dingoes. The world felt like a manga cartoon. But being on strong plant-based drugs while in the jungle added a whole new layer of complexity: was that three-headed black jaguar gibbering monstrosity for real or something that would go away after a hot bath and lots of vitamin C?

It reminded me of a spectacular hallucination experience that occurred during my undergraduate days. As university is a time for experimentation and the seeking out of new existential knowledge, a mate of mine ate a handful of psychedelic mushrooms while at the foothills of Mt. Hood. His plan was to start out at one of the picnic grounds and then have a lovely day walking in nature while wasted. What could possibly go wrong?

He had the most vivid hallucinations about conversations with a newfound leprechaun friend, singing, dancing, and laughing the day away. The next day he was checking his camera at the same time that a news bulletin came on to the television screen, stating that a missing child had been located after being lost in the woods. He gazed in horror at his camera, at the TV, and back at the camera. His "leprechaun" was actually an eight-year-old child with Down syndrome who had wandered away from a group of relatives having lunch at the same picnic area. By the time anyone noticed he was gone, he had already bumped into my friend, who was now walking, laughing, and talking, with the "leprechaun," leading him along the path to the deepest, darkest part of the forest. Once there, my friend bid his leprechaun friend goodbye, since leprechauns live in the deep forest. He then wandered happily back, leaving this confused special-needs kid in the middle of the forest. My mate debated it for a while and then deleted the photos, since they could be considered evidence. I thought it was a very smart move!

One night around midnight I went for a long walk alone under a full moon, with my torch turned off, using the moonlight to guide my way through the forest as Godsmack's song "Voodoo" echoed in my skull. I stopped to kick back against a tree. While I was there, two vampire bats started stalking me in the air, cruising silently feet above my head. As I wasn't on any sort of chemical at the time, I knew these were real, not drug-induced hallucinations. I kept really still, like sleeping mammalian potential prey, hoping they'd land and come over to try and feed on me. Along with the other vaccinations prior to leaving Melbourne, I had had my rabies shot, so I was well prepared. Sadly, they were skittish and eventually flew off. As an aside, the rabies vaccination was the most horrid vaccination I've ever gone through. It takes three shots and each time my body reacted violently with explosive diarrhea a few hours later; the third time I broke out in a full-body rash. It was like having a hypodermically delivered exorcism while being prepped for a colonoscopy.

The next morning, I was having my bath in the clean, clear jungle stream. As I got out I brushed up—luckily just with my leg—against a large fishing spider. It promptly bit me, and caused a one-inch circle of dying tissue over the next few days. I was already prophylactically crunching massive amounts of powerful broad-spectrum antibiotics to counteract the daily insults my body was receiving, knowing all too well how quickly wounds can turn septic in the tropics. Luckily in this case the tissue grew back uneventfully.

Another inhabitant of this area is the Amazonian turtle. It is a water turtle that spends an unusual amount of its time on land. This is because the sandy-bottomed jungle streams are lifeless. The only vegetation other than sparse, fine moss is the dead leaves that give the water its dilute tea appearance and taste. While this meant it was safe for us to drink the water, it also meant that it was virtually sterile. With nothing to eat in the water, these fish-eating turtles had adapted to eating nuts at the base of the trees. They would come out on land to forage for this meal like weird armored squirrels. Dick's group was also investigating another fascinating aspect of other Amazonian turtles—they seemed to communicate with each other underwater by making fart noises. This was as weird as the Australian Fitzroy River turtle, which can actually breathe through its butt!

When taking down my hammock for departure, I did not notice a scorpion that had taken up residence in one of the folds. That is, until I was stabbed on the back of my hand by the black-tipped stinger that led from the purple body. The pain was immediate, and felt like my hand was in a flame. More worrying were the effects on my heart that showed up within twenty minutes. My heart would race, pause, race, pause, race, pause, and repeat, averaging about ninety beats in thirty seconds before pausing completely for the longest five seconds of my life. Without oxygen-delivering red blood cells being pumped through my veins, I would start feeling suffocated even though I was hyperventilating from this terrifying turn of events. My balance became even worse than usual and I had extreme photosensitivity. There was nothing to do, however, but pack up the camp and head for the boat, as it was scheduled to depart with or without us. The nearest antivenom was several days away by boat, and so there was no recourse but to burn it out during the hike. Six hours later we reached the first of many boats. My hand still burnt, but my heart was settling down to a normal rhythm. I was still weak, but starting to perk up. Now, instead of feeling like death, I felt like something a cat puked up. Not an experience I'd like to go through again!

From Manaus I went on to São Paulo. I spent many happy hours wandering through the aisles of the herpetology museum located within the Instituto Butantan. This repository contained many venomous snake holotypes. It had broad implications far beyond herpetology, but I wanted nothing more than to examine the preserved body of some obscure coral snake. It was a truly magical place, the herpetological equivalent to being in a library in ancient Athens or Alexandria.

After returning to Australia, I got to work on researching the samples that I had collected. It was satisfying to discover that even small, completely harmless but still technically venomous lizards like the arboreal alligator lizard had extremely complex venoms and therefore were great sources of novel compounds with potential for drug design and development. Even more satisfying were the results of the vampire bat research. While vampire bat venom had been the subject of considerable research, the efforts had mainly been concentrated on the large clot-busting enzymes. Alejandro had even patented one of these compounds for medical use. Meanwhile, the very small components had been ignored. We discovered that one of these was extremely selective for the very small arteries of the skin, keeping them dilated and the blood flowing. Obviously, this was beneficial to the vampire bat in its feeding, but we immediately recognized it as having great potential to help re-establish blood flow for patients with skin grafts or reattached amputated limbs.

Shortly after my return, I woke to a phone call about another snakebite to yet another mate of mine. This time it was an envenomation to Myke Clarkson in Los Angeles by an obscure African burrowing snake called a stiletto snake. These unique snakes are characterized by having hinged fangs like a rattlesnake, but instead of moving on a vertical axis, they swing out sideways. This means they can erect their fangs without having to open their mouths— an advantage while attacking a potential prey item underground. Combined with a spastic, unpredictable mode of movement, it makes them among the most difficult of all snakes to work with. Myke was stabbed by one fang in his left thumb. It rapidly swelled alarmingly and became discolored to a sickly green-blue. Such spectacular local effects as his had not been well documented for this type of snake, and it had us quickly raising a lot of questions. As this snake was so taxonomically distant from any snake venom for which an antivenom was made, there was no therapeutic treatment available.

There were several immediate complications. One acute early effect was that he became delirious from the pain. Morphine had no effect, which meant the venom was acting upon a very unusual receptor or pathway. He also broke out in full body hives and had explosive diarrhea and vomiting—the same reaction I had to death adder venom, but he was having it to a painkiller. His thumb developed into a green puffy mass that strongly resembled the effects of gangrene. It was obvious that tissue was dying. Untreated, this had the potential to kill him, as it was a fertile breeding ground for all sorts of bacteria. Sick with worry, I was frequently on the phone with his wife, Rebecca, and could hear Myke howl in the background; absolutely delirious in his suffering. Rebecca was a nurse, so she understood all too well the gravity of the situation.

In addition to the zombiefication of his thumb, there was the ongoing problem of another effect of the venom. Instead of lowering the blood pressure like a Stephens' banded snake or a Komodo dragon, it raised it. We knew this was coming because what little research had been done on this venom had concentrated upon the toxins that raise the blood pressure to dangerous levels such as 220/125, which could result in an artery bursting like an overinflated tire inner tube. There was a further complication, one that shows the true domino-like action of venoms. Myke had a pre-existing heart condition. His mitral valve was prolapsed, resulting in valvular insufficiency; the valve did not open and close properly. When he was a kid this had led to several critical medical difficulties.

The decision was made to do an emergency debridement of the thumb; to cut away the dead tissue until only healthy tissue remained. This would leave Myke with significant scarring and potential loss of movement, but there was nothing else to be done. The pain he was in meant that Myke had no sleep for three days before being prepped for surgery and given general anaesthesia. When he woke, he looked at the bandaging on his hand and noticed there was less of it than there should have been. He was then told that his thumb had been amputated. The surgeon had kept cutting and cutting but found nothing but rotting tissue. Upon reaching the bone, it was discovered that it resembled Swiss cheese; it too had been damaged beyond repair by the venom. Myke was now without his most important finger. It was going to take considerable physical therapy for him to adapt to this sudden turn of events.

11
PAKISCARY

S mashed to the gills on the unstable chemical reaction between sleeping pills and alcohol was not how I had planned to arrive in the most hardcore country I had gone to yet. It had all started when the Australian company BHP Billiton began having extreme snakebite issues at their Zamzama gas fields in the Sindh desert. This is one of Pakistan's most dangerous areas, because various factions were fighting in the name of ideology. The ideologically linked areas of "Special Importance" were, by sheer coincidence I am sure, almost the exact boundaries of large natural gas deposits. They all wanted to control these cash dispensers.

Zamzama security was tight and lethally efficient. The head of security and his offsider were both former British Special Air Service officers with combat experience in areas pretty much exactly like this. While they could control the bandits, they could not control the snakes. One worker was bitten when he came out from the mosque after midday prayers, put his big workboots back on and felt a stabbing pain. A large Russell's viper had taken shelter from the sun inside one of the boots and was not happy about being disturbed. As no antivenom was stocked on-site, and the company plane was in Karachi picking up supplies, there was no alternative but the long drive to Karachi. Four high-speed hours later, while still being driven, the worker died from blood clotting issues including thoracic bleeding and intracranial haemorrhage. A second worker was killed when a Sind krait bit him in the neck while he was sleeping. He died just like my mate Joe Slowinski did: paralysis of the diaphragm, thus knocking out the ability to breathe, and leading to a death of slow suffocation. Then the site administrative manager was bailed up in a corner of his office by a six-foot black cobra. He escaped being bitten only through the quick action of the security staff, who shot the snake. This all occurred in the space of two months, when the warm monsoon rains brought the desert alive.

As the occupational health and safety issues had hit such an extreme that the site was in danger of being shut down, they contracted my mate Sean McCarthy and me through Snake Handler, the company run by Sean and his wife, Stacey. Snake Handler is the only company in Australia—perhaps the only one in the world—with certification from a higher educational regulator for occupational snake management courses. We were going to give a highly adapted version of this training to reflect the rather special set of circumstances. We don't usually have to incorporate razor wire and bomb barriers into our educational plan!

The limo and driver picked me up at 3:30 a.m. for my business class Emirates flight to Karachi, with transit through Dubai. I was the point man on the operation. I was going to arrive seventy-two hours before Sean and get everything sorted with snakes and the on-site set-up. The flight to Dubai was a long one and I had been up all night preparing the pelican cases full of vital gear, so I definitely could use the shut-eye. I was armed with a new type of sleeping pill I hadn't tried yet, one called Ambien. As I was prescribed it just the day before I left, I hadn't had a chance to dig into the literature about its specific biochemical targeting. I had settled into a traveling routine using benzodiazepines like temazepam and other valium-like drugs, and naively treated this new pill the same way: I washed two down with a double white Russian, one shot of vodka for each. This is something I'd found works a treat with benzos: if I'm tired and take that combination, I have a nice little nap and wake up serene. If I'm not too tired, I don't fall asleep and just enjoy this rather nice way to pass the time.

I woke just in time for the very tasty meal being served. Afterwards, I figured that since I had had such a lovely little nap and woken up so refreshed, I might as well repeat this chemical combination. As I was settling in ten minutes later, I heard over the loudspeaker, "Please fasten your seatbelt." "Huh?" I thought. I wondered if we were stopping in Darwin to pick up more passengers. We couldn't be much further than that. Could we? As it transpired, we were much further along in our journey. In fact, we were now arriving in Dubai. I had been unconscious the entire flight. I was quite relieved that I hadn't woken in a pool of my own urine! I now faced a two-hour layover before the two-hour flight to Karachi with a short-fused chemical time bomb ticking away inside me that was going to go off very soon.

I quickly asked the stewardess to please get me six shots of espresso. She ignored the no-serving light, probably because she wanted to see me slam them in quick succession. It kept me awake but very high and wired as I went into the Emirates transit lounge, heading straight for more espresso. I had a double shot every twenty minutes. This kept me awake long enough to make my connecting flight. I stumbled toward the boarding gate; once onboard I closed my eyes for the shortest of naps, and next thing I knew we were in Karachi. Luckily, I had a local fixer hired by BHP to meet me and ease my way through the airport bureaucracy. Very helpful, since when I tried to talk to the customs officer in response to his questions, I was so inarticulate I sounded like I had a severe head injury. I was like a concussed kitten on a ketamine trip.

Leaving the airport in a chemical stupor, at first I thought it would be a good idea to raise my arms like the returning messiah to the cheering crowd gathered outside. But my lone functioning brain cell expressed its doubt that the crowd was for me. The validity of this statement was recognized even by my gargantuan ego. The cheering crowd was there for some local politician/warlord (the difference between the two being slight, if it exists at all). The overenthusiasm of some, approaching religious rapture, to me seemed quite contrived. I concluded it was largely a rent-a-crowd to pump up the politician's fragile ego and simultaneously artificially inflate the public perception of his standing. I just kept my head down while I was quickly shuffled off to a waiting armored Mercedes. During the drive to the safe house, I was hidden behind not only the tinted windows but also the black mesh sightscreens on the inside of the back seat passenger windows.

The next day I headed off to the University of Karachi to meet Syed Ali and his PhD student, Mehtab Alam. The previous government had been led by the military strongman Pervez Musharraf, the commanding general who had seized power through a military coup d'état. While his means of obtaining power left much to be desired, as did his bloody military history, he ironically was the best thing ever to have happened to the state of Pakistani science. Much of this was due to the incredible leverage wielded by the active nuclear program that was the brainchild of Abdul Qadeer

Khan. As the University of Karachi was where Khan had his laboratories, it was the greatest beneficiary of the program.

While I had been impressed with the advanced state of some of the laboratories on campus, the legacy of this scientific investment caught me unawares. It included a countrywide Internet video–connectivity for science lectures, so that not only could someone from any part of the country view a lecture in real time, but they could also interactively ask questions of the speaker. The public seminar I gave on my venom research went in a very routine manner until we got to the question-and-answer session at the end. A twenty-something male looking like a stereotypical Middle Eastern villain from the latest unimaginative Hollywood action movie popped onto the screen: soil-stained checkered turban, scraggly beard, and close-to-gether unblinking eyes. In surprisingly lightly accented English he said, "I don't have a question, I just wanted to say that I have long followed the various updates of your webpage venomdoc.com and I just wanted to say that it is great and I have learned a lot about the snakes I also love." For once in my life I was speechless. I almost fell off the stage, I was so stunned! I stopped paying attention to where my long strides were taking me as I cruised across the stage on autopilot while digesting the shocking incongruence between the hostile visage and the very kind words.

After the talk, Syed and I headed back to his lab with Mehtab. I had made arrangements with them prior to my arrival to secure some snakes for us to use in the training and thus not have to rely on catching snakes in the gas fields. Naturally, we would still go out snake hunting while in the desert, but our task was too important to risk not having sufficient numbers of suitable snakes on hand. Waiting in the lab were four Russell's vipers and three black cobras. The Russell's vipers were light amber with black-outlined maroon oval markings down the back, while the cobras' patterns were pure indigo. Stunningly beautiful specimens of two of my all-time favorite snake types. In order to transport them to the gas field, we double-bagged them, put them into a sturdy box, and loaded them onto the back seat of a second armored Mercedes. In addition to the armed driver, another heavily armed private security agent rode in the front passenger seat, with a very large shotgun resting between his legs. My head was spinning slightly at the idea of venomous snakes being driven under armed guard across one of the world's most dangerous deserts. Even for me, this was a new one.

However, despite being in a Mercedes with bulletproof windows and metal plating protecting the rest, it was still too dangerous for me to be driven to the gas fields. I would almost certainly disappear without a trace along the way. Instead, I would be flown there. The only question was when. There was sporadic gunfire on the streets near our safe house, which made it very difficult to get to Karachi airport. After much debate, which was a bit like a conversation of "Hey, when do you think the rain will stop?" but involving bullets, the decision was made to risk the run to the airport. Back into the Mercedes, I was off to the airport, much more alert than when I arrived. I was soon whisked through departures to a private plane and on my way.

After a four-hour flight, we were at the gas fields, where we were met by four-wheel drives driven by more bodyguards, carrying machine guns this time. At the compound, as my gear was being unloaded from the vehicles, I was given a site induction. It commenced with the pointing out of the panic button and gas mask located inside by the door in my quarters. This was accompanied by a briefing on what to do in the event of an armed or chemical attack upon the compound, including where the extraction points were. I then met with the site doctor in order to review their medical protocols and newly acquired stocks of antivenom. Their occupational health and safety write-ups regarding envenomation were actually quite complete and only needed a little modification, such as adding information about the use of pressure-immobilization bandages. As all the local snakes were devastatingly potent and extremely fast-acting, it was my professional opinion that such first aid was reasonable for all species.

There is quite a debate globally among health professionals regarding the use of such bandaging. It's standard for use on Australian snakebites because the effects are almost exclusively upon the nerves and blood, with local effects being typically only minimal. For non-Australian snakes, particularly vipers, there is a great reluctance to use them, based on the logic that it might make local tissue destruction worse due to the venom being concentrated in a small area. The latter is a reasonable consideration where a species is not highly toxic and deaths are rare. This is particularly the case for snakes such as American rattlesnakes, where the local tissue damage is so severe that necrosis may set in and the affected limb require amputation.

However, I am of the view that if it is a "life versus limb" consideration, there is absolutely no question about whether a compression bandage should be used. If I were bitten by a species known to be armed with a venom that causes local tissue death and was not likely to die before I could reach a hospital and have antivenom administered, I would certainly forgo the use of pressure-immobilization first aid. But if I were bitten by a species of any type for which lethal effects might occur before proper medical help could be reached, then I would apply pressure-immobilization first aid without hesitation.

While there was much hot-air debate, there had been shockingly little actual research into whether pressure-immobilization bandaging actually worsened the local effects. Interestingly enough, one of the few studies that had been conducted was actually upon Russell's viper venom. The evidence gathered revealed that not only were the systemic effects slowed down, but the local effects were not worsened. In any case, a bite from the local snakes would be fast and lethal either through neurotoxic actions (black cobra and Sind krait) or devastation of the blood chemistry (Russell's viper and saw-scaled viper). So in my professional opinion, pressure-immobilization bandaging should be administered routinely and quickly. I discussed all these nuances with the site doctor and he assiduously made notes. Their antivenom stocks consisted of either Pakistani or Saudi Arabian products and thus, between the two, they would expect reasonable cross-reactivity with the local snake venoms.

The next day, I unpacked the snakes. They had arrived during the night while I slept lightly, bathed in the dull red glow of the panic button and with my gas mask within easy reach. As I was setting up the office that was to be our headquarters for the next week, I felt it was appropriate to have the movie *Team America* playing on the laptop, but to avoid any "misunderstandings" I kept the volume very low.

We were going to train representatives from the security and engineering teams, so I then went to meet the team leaders of these two staff divisions. The first order of business was to inspect the sites where bites had occurred, or where snakes had been reported. Unsurprisingly, the epicenters were areas of shelter, food, or water. I inspected what measures they had taken to mitigate risk. As with their medical plan and antivenom stocks, they had done a very thorough job. I was able to point out some areas in building construction where metal grates had been put in to block access to snakes, but where some erosion had occurred during the torrential rains. This was how the krait had bypassed the snake-proofed doors and entered the sleeping quarters, resulting in the death of one of the workers. Most loose material was on flat pallets raised off the ground by resting on large blocks, thus creating an unfavorable habitat for snakes. Waste disposal behind the kitchen was also an area of superb management. All food waste was double-bagged to reduce the smell signature to mice or rats, and the large waste bins were emptied into garbage skips offset from the buildings. This was still the inevitable major attractant to snakes, due to the inherent populations of rodents, but the Pakistanis had done as good a job as could be expected and certainly far and above that which I had seen at other sites across the globe.

After this very long first day, I hit the site gym and then settled down to watch cricket with some of the staff. Once they found out I was a bit of a cricket tragic, they invited me to play in the game held each evening. The matches took place under the floodlights of the guard towers, with one of the machine-gun-toting guards acting as the umpire. Obviously, dissension was nonexistent. To discourage power hitting, any smash over the razor wire security fence was given only two points, not six, and the person was out. The reason for this was that the compound was on lockdown status each night, which meant the cricket ball could not be retrieved from outside the compound until first light the next day, by which time the children from the nearby local village had already picked it up. The team I was on fielded first, so I quickly had a chance to unleash my signature style of bowling, which is as erratic as it is fast, with each run-up just as likely to result in a no-ball as a cartwheeling wicket. One hit bystander later, I had secured my team a pair of wickets. Once we were batting, as I have a natural urge to crush a cricket ball, one that cannot be denied even if I'm trying to keep the ball on the ground and run safely for points, it only took three deliveries to tempt me to step up to a spin ball and hit it sweetly. Up it flew through the moth cloud around the lights and disappeared into the darkness beyond the fence. Two and out for me.

The following day was spent finishing the site inspection and typing up the report, including required improvements to the already excellent snake-proofing of the complex. I then carefully considered a number of areas for the different aspects of training. We wanted to make it as realistic as possible, particularly for the more challenging advanced aspects, but we, of course, did not want to lose a snake in the process. After that, I repeated the previous evening's activities: workout, dinner, watching a bit of cricket on the television, then the regularly scheduled Night Test.

It was interesting to consider the social dynamics of such a pressure cooker of a remote site in a hostile environment. In line with the social norms of Pakistan, and with the majority of the workers being Pakistani Muslim, it was a male-only camp, which would be expected to lead to pent-up sexual frustration. However, this was very much not the case. In fact, there was a higher level of social benignity than I'd experienced in such situations previously. I put the strict ban on alcohol down as the major positive contributing factor. Of course, guards walking around with machine guns no doubt were also a massive contributor to the startling lack of the typical fuckwittedness that characterizes remote mining towns in Australia.

The following day Sean arrived and it was time to get down to business. First up was a series of lectures by Sean and myself regarding venomous snakes, their toxins, their natural history, and how to safely work with them. The core message was simple, but one with a deadly efficient message: it is safer to catch and remove a snake than it is to try to kill it. This was coupled with the fundamental premise that snake reduction through good planning is the ideal scenario. It is, of course, unrealistic to expect a snake-free environment, but the numbers can be minimized through the reduction of favorable habitats. Then it was on to the training.

We started the workers off on rubber snakes to familiarize them with where and how to grip a snake with the tongs without damaging it, while a team member had the hoop-bag ready for the snake to be deposited into. We emphasized the basics: the snakes are much more likely to go into a black bag, thinking it is a hole, than they are into a white bag. We also stressed the importance of double-bagging in order to minimize the odds of escape; and lastly, that the bags must be placed inside a sturdy, crush-proof, well-ventilated container. We finished the day with the participants' first experience with live, large diadem snakes, because they are a species very ready to bite, while at the same time being harmless to humans. This desensitized them to the concept of getting close to that which they fear the most—and a bit of blood always reinforces a lesson. However, this crew was the most attentive and diligent of any group Sean or I had ever worked with. They displayed a willingness to listen and the ability to learn. They also had a tunnel-vision focus upon the task at hand. There weren't any mishaps with any of these long, agile snakes—other than that a few of the crew discovered the joys of getting crapped on by a snake.

At the conclusion of the day, we joined them for a new evening activity: beach volleyball. There was a well-set-up court on the far side of the compound, also under guard tower lights. The game was far less intricate than the ones I was used to playing in my competition team in Brisbane. The players packed themselves, about twenty to a side, into a solid mass of beard. There was a conspicuous lack of bumping or setting. Rather, hits largely consisted of a unique style that looked like some sort of mutated tennis two-handed backhand. The balls rocketed back and forth at high speed, often with some unexpected spin brought about by the unorthodox hitting style. Bumping was merely a case of it hitting off someone's head at high speed. When Sean and I did a tidy bump-set-spike between the two of us, we were looked at like the white aliens we were. Quietly, and just between us, Sean and I named this new game "scud ball."

The next day we gave a brief review lecture that was a synopsis of the previous day's training and had all the teams redo the tasks they'd learned. We had them working in teams of three: one with a hook, one with a pair of tongs, and one with a bag. We had them switch equipment and roles until all were adept and the teams fluid. We then brought out the venomous snakes, first placing them on the wide lawns with shortly shorn grass. The cobras and vipers provided good examples of the two basic snake types of the region: thin, fast-moving snakes that are difficult to capture but, once captured, are easy to handle; and stout, slow-moving snakes that are easy to capture but then very difficult to handle due to their strength and long mobile fangs. We stressed a hands-off approach to snakes, and that under no circumstance should they try to pin a snake and grasp it behind the head. Emphasized continuously were the three Cs: calmness of demeanour, clear line of sight, and communication between team members.

Over the next few days we rapidly moved them on to more progressively challenging tasks, such as retrieving a cobra from under a lawnmower or a viper curled up in a large flowerpot. We then placed three large cobras inside the large metal garbage skip, and multiple vipers underneath as well as within the stacked containers of glass bottles. The final exam for each team of three was finding multiple black cobras let loose at night among the bins, gardening equipment, or rubbish. We also took them out into the field to show them the natural ecology of the snakes so that they could better understand their behavior. Whenever we left the compound, Sean and I were each assigned at least three bodyguards with machine guns to accompany us at all times. Of course, we used them as assistant snake catchers.

Among the participants was the site doctor. At the beginning, he was terrified of snakes but he gamely took the course and by the end of it was absolutely in love with the diadem snakes and posed for a picture holding one, which he then proudly displayed on his desk. That was a notably special outcome of the training. The cross-cultural communication between the group and us was among the most satisfying of my life.

Yet there was one very unsettling local interaction. As we were driving through one of the nearby small villages on the way back from snake catching, I spotted graffiti in the form of a ten-foot-long, three-foot-high multicolored flag with a sword horizontally across it and boldly written Arabic lettering above. I did not need to understand the local language to get the gist of what was being so clearly stated. I rolled the tinted window down and started snapping photos, only to be photo-bombed by a man with wild eyes who appeared from nowhere right in front of my face. He filled half the photo, with just a red motorcycle in view in the remainder of the frame. Ten minutes later, the same motorcycle caught up with us at high speed; the figure riding it, clothed from head to toe in white cloth and with familiar wild eyes, gave me a death stare. I was convinced that he was a suicide bomber and that we were going to die. The guards felt the same. They trained the machine guns on the motorcycle as we accelerated away. Had he pulled closer, they would have opened fire and filled him with holes. That was enough for me.

This was the last time I left the compound before we returned to Karachi. Back in my room, while typing up the notes for the day, the Bullet for My Valentine song "Scream Aim Fire" seemed rather appropriate to have blasting out of the speakers, as did Slayer's "War Ensemble."

Back in Karachi, we milked the cobras and Russell's vipers used in the training and Syed stored the venoms away. I expressed my gratitude for his and Mehtab's help and invited them to visit me in Australia. Once at the airport, we ran into a few complications that were trivial issues magnified by a generalized hostility toward Westerners. First was that I had forgotten to print out my itinerary, so the petty dictator at the door was not going to let me in, even though Sean had his. After much arguing, I managed to catch the eye of an Emirates agent walking past and we were ushered in. Then at the security scanning of our luggage, our snake hooks and tongs were viewed with grave suspicion. The language barrier was not helping our cause and we were getting nowhere with our explanations of their non-weapon nature. Luckily, when a more senior person was called, he looked at me and said two English words, "National Geographic?" He had recognized me from various nature documentaries on television and we had no more concerns.

They did, however, all take great interest in my Toughbook laptop computer—not out of concern, just marveling at it. So I decided to demonstrate its famed indestructibility by slamming it against the corner of a desk. The underside of the laptop made an audible crunching sound as it impacted. In my showing off, I had inadvertently discovered the computer's Achilles heel. There was a small region that, for reasons beyond me, was not heavily rein-forced and I had punched a hole into the computer, with various bits of important-looking wiring now hanging out. As we turned the corner down the hall to go to our flight, I could still hear them laughing loudly, with evident delight.

Three days later, we arrived back in Australia just in time for Christmas. On December 28, 2009, the Taliban attacked Karachi for the first time ever. A massive bomb turned the windows of our safe house into a storm of slicing triangles, illuminated into a cascade of reflected color in the dust-filled air as the harsh sunlight beamed in through jagged cracks in the wall. Forty people were killed and scores more injured.

I spent Christmas, New Year's, and all of January incapacitated with some sort of dysentery I had picked up in Pakistan. Eight pints of blood-laden diarrhea came out of me over the course of each day. I lost many pounds due to the malnutrition resulting from the inability to keep anything inside me. My doctors struggled to get on top of it. Treatment was hampered by an inability to diagnose exactly which weird microorganism was causing it. Since the doctors could not give it a name, I just ended up referring to it as Pakistass.

Once I recovered, Syed flew over and we investigated the venom samples. It turned out that the black cobra venom was extremely different from that of other cobras. While it affected the nerves as potently as any other cobra venom, it also attacked the muscles like sea snake venom would. The cheap Indian antivenom, which had flooded the market, turned out to be completely useless. Worryingly, while the Sind krait had venom effects typical of other species of krait, the Indian antivenom also did not affect it. This had huge implications, since the Pakistan-specific antivenom was in very short supply and the Indian antivenom was currently supplied as a cure. Based upon our results, Syed and Mehtab spearheaded a new initiative to develop antivenom for Pakistan that would take these results into account.

12
THIS IS SPINAL TAP

In order to stay ahead of the cane toad toxic tide, my terrestrial field research in Australia was now concentrating on the Kimberley region of Western Australia. The target animals were those most vulnerable to the looming cane toad invasion and also occupying the most pivotal research position: death adders and monitor lizards. The death adders were undescribed and only a few had ever been captured. My hypothesis was that they were the same type as found in Mt. Isa and elsewhere on the crumbling rocks of the Australian North. The Kimberley was also thick with monitor lizards of tremendous interest to lizard venom researchers.

Iwan Hendrikx and I spent our days at Lake Kununurra stalking the water monitors—beautiful lizards that have olive bodies covered with black and yellow speckles. One of us would walk to about 160 feet away from a lizard and get its attention, then just stand there looking at it. It would typically freeze, relying upon its camouflage to protect it. With its long neck coming out from its body at a forty-five-degree angle and leading to a head held horizontal to the ground, it looked like a moss-covered branch. The other one of us would sneak toward it from a direction where it could not see us, twenty-foot surf pole at the ready, while being guided by the other person as to what angle to take and where to place the noose. As we were doing this at Lake Kununurra, the fishing pole did not stand out like it did when Chris Clemente and I were catching desert yellow-spotted goannas.

One time I had the attention of a large water monitor right at the edge of the lake. It could see me, I could see it. The lizard could not see Iwan, and all I could see of him were his head and shoulders. Iwan crept toward the lizard along the water's edge, keeping a large tree between them. Suddenly, I heard a big splash. Iwan disappeared abruptly from view at the same time, so I figured he had fallen in. He came straight back up the embankment to the truck I was leaning against. I said, "The lizard's still there. What's up?" He replied with very wide eyes, "A big crocodile just attacked me!" He pulled up his pant leg and took off his boot and sock. A red line appeared on the outside of his leg and then his flesh parted like the Red Sea. It was as if a scalpel had been run down his leg, barely missing the big tendons that attach to the ankle.

The wounds could be from only one species: the freshwater crocodile. They have long, blade-like teeth for spearing fish, as opposed to the big, cone-shaped teeth of a saltwater crocodile. There were saltwater crocodiles in the nearby rivers and they occasionally walked the few hundred feet over to the lake. In fact, we had been stalking a ten-foot saltie for several days in this same spot, trying to nose it out, since it was hanging around the caravan park eyeing off the small children and dogs. But this cut was from a big, male, freshwater crocodile and we thought we knew which one: an eight-foot long-term resident male nicknamed George.

Some Swedish tourists had been feeding him for the last two weeks before they were found out and kicked out of their accommodation. Their actions had predictably created a hazard. As it was also breeding season, this meant that George was fearless and territorial—a bad combination that could easily result in him becoming a handbag.

We knew that if we reported it, he would be shot for something that was not his fault. As freshwater crocodiles do not hunt large mammals, we knew he'd calm down once breeding season passed and people stopped feeding him fish skeletons, so we decided not to take Iwan to the hospital. Instead, we would attend to it using the large medical kit we had with us. It contained all that was needed for lacerations such as this, with local anaesthetic and the means for closing a wound, including Steri-Strips, stitches, and staples. Iwan applied pressure to the wound as I drove the couple of miles back to our bungalow. I helped him inside, dosed him up with painkillers, and applied antibiotics directly to the wound, while he took the first dose of a ten-day course of Augmentin. I then closed up the laceration using a combination of Steri-Strips and duct tape. Stitches and staples would have been better from a scar-prevention perspective, but as it was an animal bite we could not seal it. If we did, we ran the risk of anaerobic bacteria flourishing, as had happened to me when I was filming Komodo dragons with Kevin Grevioux and the doctor tightly stitched up my knee with shells and even a pebble still deep in my flesh.

Iwan had been wearing high-ankle, sturdy leather boots. The crocodile's lower jaw had closed onto the boot and sole and did not penetrate his flesh there. The long teeth at the tip of its upper jaw had effortlessly sliced his flesh as they ran down his leg, but they stopped completely once they encountered the boots. If he had been barefoot, the damage would have been much more severe and almost certainly would have included severed tendons and nerve bundles—something beyond my skill to deal with. It would have resulted in a trip to the hospital and a lengthy recovery. As it was, Iwan got off very lightly, with just cool scars and a great story.

Not long after this, my friend and collaborator Stuart Parker was bitten by one of the unstudied death adders Iwan and I had collected together in Kununurra. We were keeping it at the Ballarat Wildlife Park, owned by Stuart's father, Greg. I will never forget the day he rang. From the tone of his voice when he said "Hey, mate," I could tell instantly that something was up. With us being us, my first thought was snakebite. Sure enough . . . It was particularly concerning since we had yet to investigate this venom in the laboratory, and had no idea if the antivenom would work. Luckily it did, and Stuart survived. After he was released from the hospital, he tried to describe the experience to a friend. His mate was from a *Breaking Bad* sort of area in Melbourne called Melton, and had zero experience with snakes. Stuart said, "All I knew was that I had some really bad stuff in me and the only question was how long would it take to kick in, how bad would it get, and how long would it last?" His mate digested this for a minute and then gave an understanding nod, saying in a thick bogan accent, "Yeah, I know what you mean. It sounds a lot like the first time I tried crystal meth!"

A few weeks later, back in Melbourne, I was woken with a start by the banshee screech of the smoke alarm announcing that my house had just turned into an inferno. I was having a snooze in front of the wood stove heater when I awoke to six-foot flames erupting from what used to be my roof. Something had gone wrong and the top of the chimney had set the roof on fire. I grabbed one of the fire extinguishers always on the ready near the reptile cages, rang triple-zero in an absolute panic, then emptied the other fire extinguisher. I raced up to the roof twice with buckets of water to pour on the flames. The second time, one leg went through the weakened roof, giving my leg a pretty nasty twist as I hung suspended above the floor below.

I was utterly out of my depth in a crisis for the first time in my life. I know where I stand with any venomous animal, but fire is the one thing that truly scares me. The flames were racing along the inside space between the roof and the ceiling. It had already spread from the lounge room to the computer room and the house was heating up real fast. A three-foot-thick blanket of smoke covered the ceiling all through the house. My dingoes were safe outside but absolutely out of their minds with fear.

As always, there was some unexpected humor. One of my neighbors, the only one I actually know in my part of the mountain, is also the Community Fire Association coordinator. She was away from the mountain at the time she saw the fire on her pager and realized it was the house next to hers. She rang the truck coordinator, asking them to make sure the dingoes were okay, and warning them about the electric fence and the large quantities of venomous snakes and large carnivorous lizards in the residence. A rather unusual call-out for them! The CFA crew were absolute legends. In less than five minutes, they were here in all their technicolor-lighted glory and tearing the roof apart from the inside to get to the internal fire. They put it out in no time, simultaneously rendering the house into a black swamp. Ash and extinguisher powder covered all available surfaces like a funeral shroud. I was on the phone with the insurance company while trying to calm down the absolutely freaked-out dingoes. Fire was their worst nightmare. Nothing scares an animal (or me) quite like it. The firemen checking out the lizard cages were asking, "Is there anything in there we need to worry about?" "Not really," I ambiguously responded, while thinking that the degree you need to worry about them is directly proportional to how attached you are to your fingers.

Once the fire was out and the smoke had cleared, a postmortem was done. The CFA director said that if I hadn't been so diligent about making sure I always have a fresh battery in the smoke detector, I would have certainly died of smoke inhalation. She also said that this type of fire, in between the ceiling and the roof, is the worst kind and usually the house is lost and, more often than not, they pull bodies out of the embers. Indeed, mine was the best outcome they'd ever had. Then again, they'd never been called out to a reptile keeper's house. I have two fire extinguishers always pressure-tested and good to go—one in the house and one in the detached reptile housing building.

I was cleaning up until dawn, crashed for a few hours, then continued mopping. The combination of powder from the extinguishers, ash, and water is the worst mess to clean. I was leaving in eighteen hours and an already-insane schedule had just gotten phenomenally more complicated. I had a multi-country, Odyssean voyage ahead of me. As I cleaned the house frantically and sorted the gear, the dingoes were contentedly asleep, all forgotten, not a care in the world.

Considering the blackness of my house, it seemed only fitting that I was scheduled to be a VIP at the Megadeth/Slayer double-bill the following evening. The crowd was mostly male, unshaven, tattooed sorts, so I fit right in. The few women around were wearing the latest-and-greatest fashions from *Biker Babe Weekly*. Apparently, black was this season's black. The opening act, the Australian band Double Dragon, did a workmanlike job of warming up the crowd, leading them on several drum-driven chants of *Megadeth!* and *Slayer!*

Then came Megadeth, or as they are more rightly known,

"Dave Mustaine and Three Other Guys" (perhaps they should be renamed MegaDave!). While Dave's vocal capacity spans the full range from A to B, he uses it with precision and devastating efficiency. Like the Pied Piper he led the crowd, in this case into a passionate pit. Peace might sell, but the crowd wasn't buying that. They were, however, eagerly lapping up all that he put out to tender. By the time Megadeth finished their scorcher of a set, the front of stage was a holy war indeed.

Slayer then came on and the concert, unfortunately, descended into a true season in the abyss. The first song had sparing vocals but was the hardest wall of metal I have ever had the privilege to be assaulted by, wrapping the crowd up in its muscular arms just as one of guitarist Kerry King's beloved pythons coiled around its prey. The drummer was hitting the frenetic pace that can only be fueled by enough energy drink to give an elephant a heart attack. After the first song, Tom Araya motioned for the crowd to be quiet and then informed everyone that his voice was shot and that there wasn't going to be much singing.

Slayer then launched into a series of instrumentals. Not dreadful free-form jazz like in the movie *Spinal Tap*, but something much heavier. But in the absence of vocals, the pace could not be sustained and the band flagged. The rapport with the crowd was also lost when there were complete blackouts between songs lasting a minute or longer as roadies wandered around with flashlights, perhaps looking for Tom's voice. So the band brought on random people from backstage to sing. They did it in good spirit and while the crowd were positive about the attempts, none of them had the power of Tom in full flight, when he has the voice of two mortals. The filler attempt reminded me too much of the movie *Rock Star*. I half-expected one of the walk-ons to launch into a spirited rendition of "Stand Up and Shout," including the sustained high-pitched vocal note. Slayer finished their set abruptly and without the obligatory encore. I hoped Tom's voice would hit its former glory and that this was not a harbinger of the tour to come. Nonetheless, the show was worth going to for Megadeth alone.

Backstage catching up with the equally snake-mad Kerry King, I drew for him on a beer-stained napkin a map to a good spot I knew in the mountain ranges outside Alice Springs for the gorgeous maroon-colored Bredl's pythons. We then continued our discussion of the merits of different pythons in regard to their suitability to captivity. Kerry brought up a point I hadn't considered: he said he stuck to the smallish types of carpet pythons because he was, quite rightly, worried about the tendon damage that could result from the teeth of a big species like a reticulated python. "This is why I stay well away from monitor lizards," he commented, looking at the extensive scarring on my hands. Meanwhile, a leather-clad groupie was desperately and unsuccessfully trying to get his attention, all the while with the most confused look on her face. As amusing a culture clash as could be had!

On May 15, 2010, I woke to the news that the natural history museum at Instituto Butantan in São Paulo, Brazil, had burnt down in a devastating fire. Over eighty thousand reptile specimens had been lost, including some that were the only ones of their kind ever found. Equally important collections of spiders and scorpions were also destroyed. Each collection was the largest in the world for that type of animal. Absolutely irreplaceable. This was a loss to science that defied belief. I was stunned and deeply saddened that such a preventable catastrophe had occurred. The management had long been lobbying the government for a direly needed upgrade to the museum, including a modern fire-prevention system to avoid exactly this. But short-sighted politicians preferred to squander the country's wealth on trivialities such as World Cup soccer, with corruption leeching away much of the remaining national funds. My grief was matched only by my anger.

The next day, I loaded up my Jeep Wrangler and the trailer to head up north for a film shoot about the coastal taipan for the European television channel ARTE. The trip commenced with the Red Hot Chili Peppers song "Road Trippin'" blasting from the speakers. I cut up the coast to Cairns to link up with my mate, Aaron Hopper, and the film crew from France. They were typically French in being sarcastically good-humored and delightfully disorganized. My kind of people.

Cairns had been recently clipped by yet another cyclone, so the flooding was extensive and severe. We got some great footage of naive tourists trying to drive through floodwaters, only for their two-wheel-drive rent-a-bomb to submarine nose-down into the water, with steam erupting as the engine block cracked from the water inundation. We also filmed a variety of snakes, including taipans that had been flooded out from their burrows by the stormwaters. It was an easy film shoot, and one of my most chilled out.

I had the pleasure of watching yet another group of Europeans trying to drive through flooded Australian roads. One day, the director managed to get stuck in the mud twice in the space of two hours. First by trying to drive—for reasons unknown—up a creek bed; the second time through hitting a large puddle at high speed, slamming on the brakes in a blind panic, and yanking the steering wheel side-to-side in a frenzy, resulting in her careering off the road and into the thick vegetation. Tempting fate, at nightfall she said, "I can't believe we haven't seen a kangaroo yet this trip. I've always wanted to see a kangaroo." These words were echoing in my head as Aaron accidentally clipped a kangaroo that hopped out of the bushes straight into our car, sending it spinning like a large, out-of-control furry tabletop toy back into the French crew's car behind us, where it promptly caved in the front grill. Here's your kangaroo!

After the film shoot, I cut across the Atherton Tablelands and then down the Kennedy Developmental Road, one of Australia's roughest tracks. I had prepped the Jeep specifically for this part of the trip. It had been given a full tune-up, and was freshly modified for some pretty extreme off-road driving, including a lift kit and very large tires. Just in case, I also fitted a heavy-duty winch on to the front. The track lived up to its stellar reputation. Anything that was not tied down took flight when I hit the first massive bump. My bones were clattering against each other and my brain rattling in my skull like a pebble in an aluminum can.

I tore across the outback desert I knew so well, losing myself in the terrible beauty of the desolation. I would routinely climb up termite mounds to look for diggings from small goannas in the crumbly bit at the top. Then, one fateful day, the crumbly bit did what its name implied it would do: it crumbled, sending me into free fall from almost thirteen feet above the ground. I had enough time to think, "This is going to hurt!" And it did. Badly. I landed flat on my back on a smaller mound three feet off the ground. When the mounds reach that size, they practically become reinforced concrete. It hurt in a strange, jarring, electrical sort of way. Once I got my breath back, I took a big drink of water to wash down a handful of codeine and temazepam for the pain and muscle spasms. I had a series of film shoots one after the other for the next four months, and I could not afford to be out of action. So I took advantage of my unusual level of pain tolerance, sucked it up, and carried on. I did, however, have an inkling that I might have done something very severe to my back, once it started taking me three strong painkillers just to get out of bed each morning, and many more as the day continued.

Next stop was Singapore, for more filming with my best mate Iwan. The gig was *Asia's Deadliest Snakes* for National Geographic International and for the Smithsonian Channel for the viewers in the United States. It was strange being in Asia again. It didn't take me long to be infuriated yet again by the spasmodic gas pedal pushing of the average Singapore taxi driver. But then again, the food was bloody awesome. I decided I could put up with the toxic yin long enough to have a steaming hot dish of burnt-chilli-cooked yang.

First we filmed in the National Museum of Singapore with curator Kelvin Lim. The main sequence was me successfully completing the dissection of the spectacular venom glands from the blue long-glanded coral snake. As the name implies, this snake is characterized by having very unusual venom glands that extend to over 15 percent of the length of its body. A dental-floss-thick tube runs from the fang as usual, but keeps on going instead of meeting a muscularly compressed gland behind the eye, like in normal elapid snakes. This duct runs down the neck, turning gently to end up actually inside the ribs, where it meets a gland that looks like a teardrop being sucked into a black hole. Kelvin was generously letting me dissect one of the museum's prize preserved specimens of this breathtakingly rare snake. Despite being Asia's premier museum, especially for herpetology, even they had only a handful of specimens that came in at a rate of one every decade or two. It was an extremely intricate dissection, and one that had me tunnel-visioned with fascination. I had no idea what was going to be revealed, so it was a formalin-fixed Christmas present. Then we filmed a bit at my old stomping ground, the illustrious Singapore Zoo, with the always-cranky purple shore pit viper.

From Singapore we flew down to Bali to film banded sea kraits and reticulated pythons. Being weightless again in the water felt amazing. Floating free in the netherworld, I no longer felt like Atlas with the weight of the world upon my shoulders, and the pain in my back was gone. On the way back from filming the dive, I saw two trussed-up goats being transported through congested traffic on the back of a small motorcycle. Just when I thought I had seen it all!

Back on land, my back pain was worse than before. I was not as agile as usual while in the cage of a twenty-three-foot reticulated python at the Rimba Reptile Park. This led to a very close call, in which I was almost bitten in the face.

The next day, while hanging one-handed off a rafter from the high, open-air ceiling of a traditional Indonesian house, my back was bent like a horseshoe as I had a ten-foot-long reticulated python wrapped around the other hand and arm. I felt something shift. This time the pain went up to eleven. When I got down and ran my hand along my spine, I could actually feel a slight kink in the small of my back. My friend Jon Griffin gave me some *kratom* to ease the pain. This traditional Asian plant-based remedy did not dull my senses like opioids, but alleviated the pain even more effectively.

Next stop was Penang, Malaysia, to film World War II ruins at night, showing that the bomb bunkers were infested with temple vipers. These snakes look like tapestries spun of black, green, and yellow thread—one of the prettiest snakes I have ever encountered. The crumbling concrete of the bunkers is a Hilton Hotel for rats. And wherever there is abundant prey, there are predators. Iwan and I could easily catch six per room as they nestled among the branches of the trees which had started growing in the room and out the glassless windows. The snakes would be orientated to be close to the cracks of the broken wall, using their heat-seeking pits to accurately lock on to a rat, even on a pitch-black, moonless night.

During filming earlier in the day, the director John Ruthven had said, "Okay, walk over to that big bush at the cliff edge. It'll look great in this late afternoon sunshine!" This particular bush just happened to be nicely framed for the camera, but we had been catching snakes exactly this way for the previous two hours—by looking on the sunshine side of bushes—so I had no hesitation about "faking" a scene that was just a cosmetically ideal recreation of something we had legitimately done. It was not like we were putting snakes into ice chests, to crudely induce a temperature-linked form of sedation. That sort of fakery does go on, and I don't condone it. This shot was simply to get the wide angle, and then it would be linked to a close-up we had actually taken with several wild snakes. It was all about building the perfect sequence.

So off I went to the bush and struck a hero pose I knew would silhouette nicely against the sky and clouds behind the cliff. I looked down and nailed my mark perfectly, then exclaimed, "Hey, there's a snake here!" They thought I was overacting a bit until I actually lifted up a snake at the end of my snake hook—a full-grown adult male. Less than half the size of a female, it retains the juvenile color pattern that is very different to that of the adult female. Males and juvenile females live on the thinner, smaller tree branch tips or in bushes. They are two-toned green to suit life at the branch endings—a darker green on the top, which, viewed from above, blends with a leaf's chlorophyll-rich dark top surface. The belly is a much lighter green, to blend in with a leaf being viewed from below, as the sunlight glows through greenly. The large, stoutly built females live in the tree's interior, in the darkness surrounding the larger branches. To camouflage against the pattern of fractured sunlight on the thicker branches, evolutionary selection pressure has resulted in the adult females being a lace tapestry of black, yellow, and green. The sexual and life-stage habitat partitioning has led to a fascinating morphological variation, one that I suspected would be paralleled in the venom composition.

The adaptability of their rodent prey to old buildings and the equal adaptability of these snakes, and their passive nature if left undisturbed, have led to these snakes being central to local indigenous religions. The Snake Temple in Penang was once one of the few buildings carved out of the jungle. The building attracted rodents, which in turn attracted the snakes. The nature-loving Eastern religion, a Taoism variant, worshipped the snakes as symbols for the beauty and power of nature. Over time, Penang has become very built up, with the snakes being persecuted and killed as their habitat is destroyed. Yet the snakes persist in the gardens of the Snake Temple, living their lives out on the grounds, with the population replenished by the annual birth of two-tone green, big-headed live young.

Next stop—getting muddy and sweaty in the Cameroon Highlands of Malaysia. When I milked an almost thirteen-foot-long king cobra, the pumpkin-orange head leading to a golden body with reticulated silver markings, I chalked it up as one of my all-time favorite field moments. I had long had a soft spot for these regal snakes, which are so remarkably intelligent. Like the Komodo dragon, king cobras are very long-lived and the biggest of their kind. They hunt by sight, using their excellent vision to spot rat snakes, their favorite prey items. Their ability to process visual information and respond accordingly inevitably leads to some sort of cold, predatory intelligence. They are relatively lithe for their length, being a fraction of the width of the massive reticulated pythons with which they share the forest. The gushing yellow liquid spurting from the short but strong fangs tells of the immense venom yield of these snakes, being more than equal to that of any other snake in the world, even the Australian champion in this regard, the mulga snake. The king cobra also rivals in venom yield the snake with the longest fangs in the world—the Gaboon viper in Africa, whose intricate geometric pastel pattern allows it to hide in plain sight among the dead leaves on the forest floor.

Then I achieved what I have to regard as my absolute pinnacle adrenaline moment. This included all previous moments, not just those that came while working with extremely dangerous venomous snakes, or even other venomous animals. It also topped other moments of lunacy, such as suicidally trying to surf big storm waves and almost ending up as the concluding scene in the movie *Point Break*. Those were all deeply satisfying and totally awesome adrenaline fixes, but this was something special: a live blue long-glanded coral snake. Something I had been after for over fifteen years. It was in the lush Malaysian rainforest of the Cameroon Highlands that Iwan and I finally came across one. We discovered it in the most incongruent manner: by zipping around a golf course in the early morning following a drenching storm the night before. We had targeted the golf course as it bordered intact rainforest, with the artificial lakes being a natural draw for all animals. This snake looked like no other. The preservative that made the tissue of the museum specimen turn to rubber had also washed out all the coloring. This snake was striped, with deep ocean blue intervening between electric blue and artery red. It was living color that also moved like electricity. My experience counted for naught, as this snake moved like it was a venom-tipped garden hose writhing from a strong jet of water. Despite being thin, it was a big snake. It gave the impression of a big snake cut in half lengthwise—the head was surprisingly long for the narrowness of the snake, and the body was deeper than it was wide.

This snake species was evolutionarily selected to be the apex predator in the specialized habitat that was the floor of the tropical rainforests. In the thick leaf litter, this animal reigned supreme. Even kraits lived in mortal terror of it. If you were a snake of any kind, you were food. It would eat snakes its own length, including juvenile king cobras. Rather than the straight-down fangs of a cobra, the fangs of this species were at a forty-five-degree angle and therefore much longer than I expected. It also moved like it was a sea snake underwater. Russian ballerinas would love this level of agility and body control. It was at the femme-fatale level of deadly beautiful. As nasty as a krait on a bad hair day—but every day was a bad hair day for this snake.

One close call was particularly tachycardia-inducing: when it struck from within the bag and almost got me. Fangs poked out of the bag, perfectly in line with my hand only an inch away, the head pushing the bag frighteningly close to my finger. In my research into what little was known about this species, I came across only three bite reports but two of the people bitten died and the survivor had his nerves torn apart in a really strange way. When feeding, these snakes need to immobilize their prey very quickly. And since it is the predator's predator, it feeds on some snakes that are themselves adept predators of other snakes. On the forest floor, an agile, fast-moving snake could easily escape, so there is an extreme selection pressure acting upon the venom that is different from that exerted on any other snake. I had long sought this snake, as I viewed it as the ultimate venomous animal, and I could not wait to get stuck into the venom research. But I still had a long trip ahead of me, with months on the road before I was back in the lab.

By the time the film shoot was ready to fly to India, I was back to crunching painkillers like they were candy. The final leg of filming concentrated on Indian king cobras, kraits, and Russell's vipers. James Haberfeld and Chip Cochran joined me on an epic quest across India, accompanied by Gerry Martin, Gowri Shankar, and Rom "The Dude" Whitaker. We commenced filming with the Irula snake tribe that Rom was working with to establish a co-op to supply essential venom for the production of life-saving antivenom. Irulas are an ethnic group of India. Traditionally, the main occupation of the Irulas has been snake- and rat-catching. These incredible aborigines had an unparalleled ability to find snakes based on even the subtlest of signs. We would come to one hole in the ground that looked just like any other hole in the ground. Not only would they say with absolute authority that there was a snake in it, but they would follow this with a statement of certainty as to what the species was. Russell's viper. Check. Cobra. Check. Krait. Check. Rat snake. Check. One after another was spot-on. I was simultaneously humbled and awed by their seemingly supernatural abilities.

From there, we traveled to the Agumbe Rainforest Research Station in the Western Ghats region, a place as famous for its king cobras as it is infamous for its leeches. The same day that we did the most epic call-out to catch and remove a massive king cobra from a local residence, I had woken up with a leech on my eyeball. At least I wasn't the one who got one on the scrotum while wading across a creek. For the rest of the day, as his white shorts turned conspicuously red, he was constantly greeted with "Congratulations! You are a woman now!"

During the time filming in India, there was one particularly shocking moment involving a local Green Cross snake catcher named Robinson. He had been bitten three weeks prior by a spectacled cobra, treated with antivenom to neutralize the neurotoxins and then released with a heavily bandaged hand. The film company wanted me to discuss with him the circumstances of his bite and to examine his wound. However, the same bandages had been on for three weeks and he had not had a single follow-up examination. I was deeply concerned about this and said to the film company that if we were going to film his case, then we had an ethical obligation to get involved with his care. To their credit, they readily agreed to pay for a visit to a private hospital.

As the bandages were unwound, the rank smell of decaying flesh made itself evident to even my snakebite-damaged sense of smell. Once the final bandage was taken off, a horror show was on display. There were two pronounced spots of extremely necrotic tissue that were displaying the insidious signs of gangrene. They were the green color of rotten cheese and smelled worse than a dead possum I once found in my pool filter upon returning from a long trip. It was absolutely revolting and extremely concerning. Untreated, such deep-seated infection could result in limb loss or even death. I was absolutely aghast at the shockingly poor treatment he had received and was quite angry that things had reached such a dire state when they could easily have been prevented. The necrosis and accompanying infection necessitated aggressive tissue debridement, with the dead material being cut out. By the time the surgeon was done, we could see exposed tendons and bone. But Robinson would keep his hand. In order to make sure that he received the proper follow-up care, the entire crew donated whatever money we had with us. A small stack of bills of all different sorts of currency was passed to Gerry, who graciously took control of it to ensure that the medical expenses were covered over the coming weeks and months. Once again, this reminded all of us what a dangerous game we played.

By the time we were done with this very long shoot, we had accumulated the footage that would ensure that it could be cut together to produce the documentary I had always wanted to make. It would be the most comprehensive, educational-yet-thrilling one of my career. Like all great career achievements, this one came at a considerable personal cost. By the time we finished filming, I was well and truly done for. With my back the way it was, getting upstairs was painful enough, let alone running across a marsh after a big male king cobra, as I had done in Agumbe. I took Gerry up on his generous offer to accompany him to the eco-station on Andaman Island. Once there, I just curled up in a hammock, smoked lots of Indian hash for the pain, and didn't move for two weeks, while soaking up the healing atmosphere. My back muscles relaxed and so did my brain.

Semi-mobile again, I was off to film Komodo dragons for a BBC *Natural World* special on my research into their venom, called "Secrets of the Dragon." I was taking child-like delight in this, since BBC *Natural World* produces the highest-quality natural-history films. Solidly based in fact, in fine British tradition, with David Attenborough the bedrock. To have that production company doing a special on my Komodo dragon venom research was a huge validation after such a long road. I was also particularly pleased since I knew they would be interviewing David Attenborough for it, with him reflecting upon his own travels to Komodo Island to launch his film career decades before. I was humbled that a man whom I had admired for so many years was aware of my research on an animal we both found to be absolutely magical.

The BBC team was coordinated by director and producer Stephen Dunleavy and cameraman Gavin Thurston, both of whom had worked with David Attenborough on several occasions, with Gavin being one of Attenborough's favorite cameramen for many years. Iwan and Sean McCarthy were coming along too. On seeing me again, both were visibly shocked at my state. I was losing condition fast, due to the back injury. As my core muscles lost tone and mass, I was less able to support my spine and felt the injury more and more. By now I was tossing back up to a dozen Vicodin pills a day just to keep from trying to kill myself by smashing a hammer repeatedly into my skull. Most worryingly, I was starting to have episodes of tingling and weakness in both legs. This accentuated the already-dangerous situation, as I didn't have any of my usual speed or agility. Luckily, Iwan and Sean were there to help with animal management and do the heavy lifting of the gear on Rinca Island. The filming went smoothly and uneventfully. Uneventfully, that is, except for the blinding pain that accompanied any twisting motion of my torso and the stumbling gait I assumed whenever the weird electrical tingling and numbness traveled down my legs. Hardly ideal for working with the world's largest carnivorous lizard.

There was one event that did take my mind off things for a while. We acquired a mascot on the last day of filming. I saw a strange splashing in the distance while we were having breakfast on the top deck of the boat. Curious as to what it was, I took off in one of the dinghies attached to the main boat. Getting closer, I could see that it was an orange-headed fruit bat. I surmised it had fallen into the water while returning to the mangrove island that was half a mile away, after a long night of foraging. It must have been too exhausted to continue. So near, yet so far. The three-foot wingspan was inefficient for swimming and it was unable to take off from the water. Without help, it would certainly die.

There are dangers in handling fruit bats, since they harbor many types of viruses. In other parts of the world, rabies would be the main concern, as with the vampire bats I worked with in Mexico. In Australasia, the rabies-related lyssavirus is the worry. As I was immunized against rabies, I was equally protected against lyssavirus. In any case, I could not have let the bat drown. I took off my T-shirt, spread it between my two hands, leaned overboard, and wrapped the bat snugly in the shirt. On returning to the main boat, I took the bat up to the top deck and let it hang in one of the unused sails of the large, old, wooden yacht we were using this trip as our floating base camp. Fred, as we named him, hung around for several hours, resting and regaining his strength, before he flew with uncertain beats of his leathery wings back to his island home.

After the field filming, we were back to the Rimba Reptile Park to see my favorite animal in the world: Monty the Magic Dragon. I would like to think that there was some recognition in his eyes at the sight of me, leg intact this time, but back a wreck. Today we were going to milk him for his venom to demonstrate its powerful effect upon blood clotting. To do this, Sean rested his large frame on Monty's back, while Iwan gently but firmly restrained his head. I ran my index finger along the outside of the lower jaw, from the rear to the front, depressing the hollow macaroni-like gland to squeeze out the liquid venom contained within. I watched it pool around the teeth as it came out and then quickly suctioned it up with a large pipette tip. Then came the fun part. My mate Jon Griffin used his phlebotomy skills, which any train-spotting junkie would love to have: he inserted a large-bore needle into the blue vein of my bulging left forearm. He drew 6 cc of my blood, which we quickly transferred into tubes. There were three each of two different types of tubes: the first each contained less than a teaspoon of venom, and the second set was the control, containing less than a teaspoon of pure water. One cc of blood went into each of the six tubes. And then we waited. Twenty minutes passed and the tubes containing blood and water were turned upside down. The blood was now like red jelly, semi-solid and immobile. The tubes containing the venom were still completely liquid— the venom had destroyed the ability of the blood to form a clot. I said to the camera: "This is my blood; this is my blood on venom. Any questions?"

The last leg of this very long and chaotic trip was a flight to Europe and the filming of the magnetic resonance imaging of Komodo dragon heads at the Leiden University Medical Center, to show their large and intricate venom glands. This was the last sequence for the BBC shoot; it went routinely and we wrapped up. I took some downtime to relax at my favorite Leiden coffee shop, the Coffeeshop Leidse Plein, but even marijuana was not alleviating my back pain.

I managed to pull myself together enough to be able to pop over to the United Kingdom with Iwan for the opening by my mate Luke Yeomans of his sanctuary devoted to king cobras. These magnificent creatures were in sharp decline in the wild. Areas I had been to years before had been destroyed by senseless logging. Whenever a habitat is disturbed to that degree, the first animals to disappear are large, slow-moving ones like king cobras. Luke's heart was definitely in the right place. He wanted to demystify these misunderstood animals; to show them as intelligent, almost sentient, beings deserving of our utmost respect. To do this, he utilized a hands-on approach, forgoing the safety equipment (such as hooks or tongs) routinely used in my expeditions. While I appreciated the sentiment, the closeness gave me a spooky feeling of an inevitable disaster. I was far from alone in feeling such dark prescience.

I then traveled by rail to Germany to check up on my collaborative research with my friend Dessi Georgieva. I took a circuitous path to get there, stopping off in Zurich on the way. There, for the pain, I was given a 50 cc bottle of a liquid opium derivative and told that it was extremely powerful and to be very sparing in its use—no more than 0.5 cc at any given time. "But of course," I blithely replied, while thinking to myself that if mere mortals were allowed but half a cc, I could tolerate a bit more. So I drank about 5 cc one afternoon in a Turkish restaurant, while in absolute agony.

As I gracelessly slid down the booth and on to the floor, with my body giving a limp sort of spasm, I thought to myself, "Hmmm . . . so this is what a drug overdose feels like. I thought it would be a lot more fun. It is exactly nothing like I thought it would be." I bypassed fun entirely and went straight to extreme sweating and nausea. I had dosed myself well and truly. The staff quickly helped me up. I gave an honest account of what had happened and said I should be fine if I could just keep moving for about thirty minutes. As I had eaten at this restaurant several times in the preceding week, two of the staff had struck up a friendship with me. After the strongest Turkish coffee they could make in a hurry, two of them took me on a little speed walk around the block a few times until I could stagger under my own steam, and could continue forcing myself to be bipedal. The similarity to the premise for the movie *Weekend at Bernie's* was not lost on me.

The next day my travel insurance had me on an emergency flight home. I was routed through Los Angeles and was to catch a flight on from there after resting for a day or two. Boarding a plane in a wheelchair was not how I'd envisioned things going, but here I was, partially paralyzed in both legs. After the flight from hell, my arrival in Los Angeles was followed by emergency transport to the Olympia Hospital, since things had degenerated from bad to worse during the flight. As my spinal cord was being pressed against the bone by one of the vertebra, at times my entire world would be consumed by pain and I would experience the purest of tortures. Acid-rain tears of frustration burnt down my face.

I soon was intimately acquainted with why this hospital was the favorite of Los Angeles' rich and famous. I was quickly admitted, once the seriousness of my condition was conveyed. This was rapidly followed by an intravenous injection that caused the clouds to part and silent lucidity to fill my soul. There is but one true god and hydromorphone art thy name. Less than ten minutes later, while I was wrapped up in this great big chemical hug, another nurse came by and said, "Oh, you poor thing. Such a terrible injury. Here, let me give you a shot." I tried to say, "It's okay, I've already had one. Thank you, though. I'm feeling much better." But she was faster, and all that came out as this second wave of heaven washed over me was "Ohhh yeeeeeaaaaaaaaahhh." Take the best orgasm you've ever had. Multiply it by a thousand, and you're still nowhere near this divine feeling.

Like Icarus, I had crashed back to earth after striving to ascend too high, flying too close to the sun. Subsequent magnetic resonance imaging and X-rays revealed that the spinal region of L4, L5, and S1 looked like a train wreck. The consulting Beverly Hills neurosurgeon Dr. Justin Paquette took one look at the results and slapped a no-fly sticker on me. As it happens, I had been traversing the globe these last four months with three completely obliterated discs and two fractures to my spine. Apparently if at any time over the last few months I had taken a hard fall, there was a very high probability that I could have ended up permanently paralyzed. Perhaps having a very high level of pain tolerance was not such a good thing after all. Then again, my extremely casual approach to self-medication with prescription painkillers was not such a bright idea either; I do have a rather addictive personality. By now, I was well and truly hooked on the pills. More direly, I had 35 percent paralysis in my right leg and 20 percent in my left. Which might, or might not, be repairable by surgery. There was no way of telling until they got in and starting cutting and that would not happen until a complex web of international insurance was navigated.

Basically, I was as fucked up as a biology science fair project at the Young Earth Creationist High School.

In the meantime, I was parked in a hospital bed. Whenever I woke up, a ghost whisper would scratch across my soul: welcome to hell. Then the drug shot into my vein would hit me: a cold fire, delightful sort of pain. Immobilizing me, almost making me think I was dead. But I wanted a chance to use every drop of ambition still left in me. Sometimes I told myself, "Just let it breathe." It was a calmness I'd long been searching for. Slipping down, so very elegantly wasted, under waves of molten amber, I would float weightless in a timeless netherworld, neither awake nor asleep; existing on another plane entirely.

Lying in the hospital bed, I wasn't able to run a razor across my skull to maintain the skin job I had had for a hairdo for the last fifteen years. As my hair grew out, it was revealed that sometime during this last decade of decadence, my hair had started to turn grey. The first few had staked out territory around the temples, the vanguards of an inevitable invasion that would spread like cane toads. Combined with being bedridden with a broken back, this inevitably led to thoughts of my own transience and the certainty of the end of my mortal days. How we live our life is not how we end our life. Do I wait for the long, whimpering end, or do I seek out the blaze of glory? Do I just keep going as I am and leave my body to science fiction when I am done ruining it? I thought of the irony that no one is as careful as the elderly, when they have the least to lose, and no one is as careless as the young, when they have the most to lose.

The essential existential angst within us all is the knowledge of our own mortality. The betrayal by the flesh. The ultimate horror is the realization of a finite existence—something everyone must come to terms with in their own way. It is only by truly accepting mortality that a life can be fulfilled. Only then are great works possible through the pouring of the life-force into a new vessel. This is the motivation fundamental to all scientists—to make a discovery and ensure this knowledge is perpetuated beyond the life of the discoverer. Immortality achieved through discovery. It gave me great comfort knowing that my scientific legacy would persist long after my flesh.

We all want to hold on to the essence of youth, and this is, of course, the reason for the endurance of the vampire legends and other tales of immortality. But in a terminal, mortal existence, a less-intellectual form of immortality can be achieved through a different form of creation: that of life. Knowing that the genetic legacy continues with the perpetuation of the lineage and pedigree. To this end, I have already also achieved immortality by selling my sperm during my undergraduate years and thus having nineteen demon-spawn. But I would be an anonymous forefather, not someone of note in a family tree.

All my life I had been obsessed with venom in all its forms—however great or small. This obsession had been the fuel driving my ambition. It had come at no small physical cost, though, during my travels to over forty countries, including twenty-six snakebites and twenty-three broken bones incurred during a life lived in no small part within hospitals across the globe. Some of the ailments had been rather mundane, with some of the broken bones and stitches the result of me losing my balance at the wrong time as part of the permanent effects of my childhood spinal meningitis, but others were quite esoteric. I had long ago become used to waking up in a hospital bed as a teaching case, with a group of medical students viewing me as some kind of weird bug. But it also came at no small emotional cost, as I had missed out on everything else while living a Peter Pan life. I was lying in a hospital bed without a wife, kids, or even any hobbies to speak of. Did I want these things? I wasn't sure. But I wanted the opportunity to find out. Is it truly better to burn out than to fade away? What about a third option? I reckoned I had a second book's worth of adventures in me, but I needed to survive this in order to write that story.

In the meantime, each day stretched monotonously into another. When not zonked from the medication, I watched lots of television. I would start the morning with *Metal Mayhem* and then *That Metal Show*. It seemed rather fitting that Ozzy Osbourne's latest video for the song "Scream" was on heavy rotation, since I'd start each day wanting to shriek to the heavens in frustration.

Once a bit of metal music had helped with my existential angst, I'd switch to watching random episodes of *Friends*. I was watching them on Direct TV, downloading from three different channels that were simultaneously running different years of the series. I'd pass out from the pain medications halfway through some random episode and wake to the year before, or years later. Matthew Perry's weight would vary dramatically as the series skip-roped its way across his different stints in rehab for Vicodin addiction that was the result of pain treatment for his chronic pancreatitis. One minute Chandler Bing would be bloated; I'd take one very long blink and then he'd be looking drawn and haggard. That kept me entertained for a while, but then I started going out of my brain with boredom and frustration.

One bedridden day, while listening to Judas Priest's "Painkiller," I realized I'd had enough of wasting time doing nothing but being stoned on prescription opioids. So I started work on a book, going into exhaustive detail about the evolution of reptile venoms, the pathophysiology of envenomations, and the drug development potential of isolated toxins. But even this was not enough to keep me in bed while in a holding pattern for surgery.

My stubbornness had me out of bed on two occasions. Once

was to give a pre-planned speech at Loma Linda University at the kind invitation of my mate Dr. Bill Hayes. Our research into Southern Pacific rattlesnake venom was driven by our desire to provide insights into the increasing fatalities from one of North America's most dangerous snakes. Sensationalized news articles had claimed that their venom had undergone recent, rapid change, leading to unusual and highly toxic effects in patients which had left clinicians at a loss. My friend Sean Bush was at the vanguard as he was an emergency physician in the bite epicenter. He had observed more aberrant cases than anyone I knew. Our work showed that rather than changing rapidly, the venom varies dramatically between different populations of the snakes due to long-term adaptation to different environments.

These snakes live in habitats as diverse as isolated Catalina Island, the high-altitude San Jacinto Mountains, the grassy hills of Loma Linda, and the desert transition zone of Phelan. In a single hour's drive from the desert floor to the top of the San Jacinto Mountains, the venom goes from destroying the blood to frying the nerves instead. Millions of years living in these very different habitats has led to specialized venom chemistry which needs to be understood to effectively treat snakebite patients. We hoped this study would help clinicians to treat bites that have had baffling effects on patients, as well as contributing to the knowledge of venom evolution.

Unfortunately, the sole American antivenom for this species, CroFab, has had poor effects in many cases, with patients dying despite receiving high doses of it. Sean had one patient die despite receiving sixty ampoules of this antivenom. There is emerging evidence that the Mexican antivenom, Antivipmyn, actually performs very well against the Southern Pacific rattlesnake venom, but the manufacturers of these two antivenoms are locked in a court battle over sales rights in the United States. This is a case of commercial interests being put above patient needs and it is leading to the loss of lives.

It was clear that reports of unusual effects were due to better record keeping and reporting, and changes in human behavior, rather than any recent changes to the venom. New housing estates are being built in what used to be remote areas, and people with low snake awareness are coming into close contact with Southern Pacific rattlesnakes. Many times when a patient presents at the hospital it is because they tried to kill the snake and were bitten in the process. I have a simple philosophy: if a person has enough time to go into their garage, get a shovel, try to kill the snake, and get bitten in the process, they have no right to throw themselves on their knees, raise their arms to the heavens and plaintively cry out, "Why meeeeeeeeeee?!" Because you didn't call a registered snake removal service, that's why. You tool.

The drive from LA to Loma Linda had me in so much pain that I was crunching a triple dose of Vicodin. My talk went okay. It was not the first time I had given a lecture heavily under the influence of strong drugs, so I was able to do a serviceable job. But it took a mighty toll. At the after-talk dinner, the dozen Vicodin I had taken in the previous four hours got their revenge. Vicodin in large enough doses triggers extreme nausea in me and I vomited harder than ever before. Even more than the time during my undergraduate years when I stage dived at a Soundgarden concert, landing on someone's head with my stomach, and two pitchers of Hefeweizen beer were instantly expelled onto the poor bastard. Only this time, spaghetti streamed out of my nostrils like a pair of mutant octopuses. Wrong meal to have before projectile vomiting out of mouth and nose.

After this little adventure, I quickly cycled through all the available pill painkillers. My pain ate them up for breakfast. My doctors finally put me back on to the intravenous big guns: hydromorphone—the sort of painkiller that stands on a rating scale of "this one goes to eleven." It got a handle on the pain most of the time. However, at least once a day there was a time when the pain was so great that I tasted madness. It was always hanging over me like the sword of Damocles. It could come from something as simple as turning the wrong way in bed. I would become absolutely consumed with pain for up to an hour, or even longer. This was also when my legs would fail completely, as it was when the pressure was greatest on the nerves. If someone had told me at that point that this would be the quality of life I would have for the remainder of my days, I would have asked for a gun so that I could euthanize myself. This was no kind of life. This was true torture, on a scale that the Marquis de Sade would have envied.

Hydromorphone gave me such horrendous constipation that it was like trying to crap out a brick. I had to eat massive amounts of prunes to try and get things moving inside my intestines, but hydromorphone also killed my appetite. Medical marijuana gave me some appetite, but it mixed with the hydromorphone in a very unpleasant way and I stopped taking it. I would get auditory and visual hallucinations, and not the fun ones. More like the time I was bitten by the canebrake rattlesnake when I was a teenager, and less like the times doing psychedelic mushrooms in the Netherlands, or the indigenous hallucination ritual in the Amazon. The combination also produced some pretty nasty Parkinson's-syndrome-like tremor effects. I was done with marijuana and back with the death of my appetite. As weeks passed, I was steadily losing weight until I eventually started looking skeletal and feeling extremely weak.

The second time I was out of bed was on November 30, 2010,

to see Roger Waters perform *The Wall* at the Staples Center for my fortieth birthday. While I suspect I was not the only one at the concert extremely drugged on opioids, I suspect I was one of the very few doing so legally while wearing a very intricate, rigid back brace. I must say, this particular class of chemical certainly added to the experience. "Comfortably Numb" took on a whole new meaning. I was in a very good mood during the concert. Just as well, since one cannot listen to Pink Floyd while depressed without killing oneself. It is not just the drugs talking when I say it was far and away the most beautiful concert I had seen. I was then banished to bed under strict orders not to leave again. Surgery was scheduled for December 23. Two days before Christmas. Hopefully Santa would bring me a new spine and working legs!

I have never been so scared as when I was wheeled into the surgery theater. Cold terror gripped me. The anaesthesiologist told me it would be a little while before they started. "Would you like to read anything while waiting or would you prefer to be put out early?" Without hesitation, I asked to have my ass knocked out right away.

Dr. Paquette was a one-of-a-kind specialist, and did a magnificent job repairing my wrecked spine. If it had been done in Australia, they would have just had to lay down the concrete and fuse multiple vertebrae. That would have ruined my fieldwork forever. He was able to perform a procedure not available in Australia at the time, whereby he put cobalt-alloy end caps on three of the vertebrae, with artificial discs in between—the same procedure he was filmed performing on Schwarzenegger's stunt double for a documentary. He also had a contract to repair the compressed backs of NFL running backs—obviously the caliber of person I wanted working on my precious spine!

Now that the surgery had been successfully undertaken and I

had a Wolverine spine, the long and painful rehabilitation began. This included having to learn how to walk for the third time in my life—an extremely painful exercise. The neurons that hadn't been in action for months were firing and atrophied leg muscles needed to be rebuilt and restretched. Being a former competitive swimmer, I was used to having to suck it up for training, but taking those first few steps was the most grueling and painful workout I had ever suffered. My sensory perception was warped. If I closed my eyes and described, without looking, the position of my right foot, I always estimated it as pointing much further outwards than it was. This meant that I was unconsciously walking with it pointing inward, in a pigeon-toed fashion. I had to concentrate on this and always force it out further than I felt it should be. If I got tired, it would steadily change position until it was pointing inward while I walked again.

Three weeks after surgery I filmed most of the episodes for season one of *Monster Bug Wars* while propped up on a special orthopaedic cushion arrangement. I remember none of it. In the footage my eyes are electric blue and my pupils are the size of pinpricks. Watching the show much later on, I concluded that the opioids were definitely a performance-enhancing substance since I was very much in sync with the rather out-there premise of the show, which was basically celebrity death matches between different arthropods. Considering how smashed the host was during filming, it was quite fitting that this show was eventually voted online as one of the best stoner shows ever.

Then it was time to take myself off the massive doses of hydromorphone I had been taking to remain sane in the last few months. I made a judgment call as to when I thought I could deal with the physical pain without the pills, and one day went cold turkey off this intravenously injected, long-acting opioid. No stepping down to a short-acting opioid and then tapering off. I felt I would be more likely to do it by taking a brutally abrupt approach—the chemical equivalent of jumping into icy cold water rather than going in slowly. While long-acting opioids do not come with the insidious cravings that the short-acting form do, there is no free lunch. The lack of acute addiction is paid for with dramatically more intense withdrawal effects.

I experienced agony and sickness like nothing I had ever felt before, and hope never to feel again. I bit down on towels to keep from screaming. Every neuron was raw and firing uncontrollably. My bones felt like they were rubbing together and even my freshly grown hair hurt. I understood now what heroin addicts say about not being able to stop despite desperately wanting to, because of the withdrawal sickness. It is just too much for them to take. I drenched the mattress due to the gallons of sweat bucketing out of my pores each day. Gradually I started feeling better and the mattress had to be thrown away since it was so saturated with my sickly-sweet-smelling sweat.

While killing time in the hospital, one day I saw a news report about biblical levels of rain and flooding occurring in Brisbane. The main river going through the city had busted its banks and large portions of the city were underwater. Skyscrapers poked out of the water like the world's largest flood posts. There was even footage of bull sharks going into butcher shops to feed on the rotting meat. To add to the surrealism of it all, I received an email saying that my latest research fellowship application to the Australian Research Council, for a Future Fellowship this time, had been successful. This meant that once I was released from the hospital and was able to travel back to Australia, I was moving back up to the University of Queensland in Brisbane.

13
PHOENIX RISING

As I presented my travel documents to return to Australia, Slash's song "Back from Cali" was playing on the iPod. The lyrics were quite appropriate. I was definitely tired and broken. I had certainly lost my way. I glanced in disbelief at the news on my phone. Just bloody peachy. Another massive cyclone was making its way over from Fiji toward Australia. Cyclone Yasi was the name given to this monstrosity. This beast was even bigger than Cyclone Larry five years earlier. Only this time, instead of being chased by a cyclone while on a boat in the Coral Sea, I was about to board a plane whose flight path was on a collision course.

The flight over was a turbulent roller-coaster—exactly what my newly reconstructed spine didn't need. I was traveling without painkillers of any sort; I had thrown away the surplus after the hellish withdrawal experience. I just had to deal with it as best I could. I was doing fine until the entire plane started shuddering like an epileptic having a grand mal seizure and the overhead bins vomited their contents. My back screamed like the tortured steel that it was. This ended up being the worst flight of my life, even worse than the medevac flight from Germany to LA. It also took longer than usual because a long, looping detour south of New Zealand had to be taken to avoid the worst of this epic storm. I arrived in Melbourne looking, smelling, and feeling like something found in a gutter.

Things had gotten a bit out of control at home during my eight month absence. The weeds reached nearly six feet, the pool was baby-crap green, and my Jeep Wrangler was so covered with dust it was a tan-orange color instead of fire-engine red. While dingoes were legal to keep as pets in Victoria, Queensland was a redneck wonderland and regarded them as pests. Consequently I had to give them up before leaving so with alacrity I rehoused them. Seventy-two hours after landing, I had the Jeep washed, the trailer packed, and was on the road to Brisbane, where I had found a house on Mt. Glorious. Rather than taking the shorter route along the east coast, I cut up through the outback, sticking to remote dusty red roads and driving at a leisurely pace. I drove only four to five hours a day before I was uncharacteristically exhausted and had to pull over and pitch a tent in the scrub. It took me eight days, instead of the usual two, to get there.

Once I dropped down the Great Dividing Range, the damage from recent flooding quickly became apparent. Houses smashed up against trees, overturned cars half submerged in putrid mud. Divots out of road edges like a giant shark had taken bites in a feeding frenzy. Rubble strewn across what was left of the road. If I hadn't been driving a Jeep specifically modified for fairly extreme off-road adventures, some areas would have been impassable. As I drove up the side of Mt. Glorious, I felt it was a good omen when, only a few miles from my new house, I came across a particularly high-contrast Stephens' banded snake. It was very unusual to see one out during the day, so I concluded it must have been displaced by the flooding. I had no need to try to catch it since I had finished with that area of research long ago. I just placidly watched it as it crawled across the road, giving me a pugnacious sideways look.

I pulled into the rainforest that was so familiar to me, with the colorful birds screeching greetings in their inimitable way. I was in my natural home again; so much had happened in the eleven years I had been away. I pulled up to my new house, opened the door, and took a breath. The funk of forty thousand years hit me like a moss-covered sledgehammer. Due to the incessant rain and hot temperatures, the house had grown green stubble over every surface. My furniture would be delivered the next day, and I had no choice but to spend the remainder of the day and much of the evening scrubbing, doing more cleaning than I had done in my entire life up to that point. Cumulatively. Next day, the furniture delivery crew asked me where I wanted everything set up. I had them put my bed in the middle of the living room by the wood stove heater, so that I could stay nice and toasty on the cold mountain nights later in the year. In the three bedrooms I had them set up the steel shelves to hold my mountains of field gear, surfboards, and my clothes in clear plastic bins—a home decor strategy that just screamed "A single guy lives here."

It was only twelve hours before I met my new neighbor; I heard screams and walked outside to see what was going on. A little old lady was trying to remove a ten-foot python from her yard. It turned out she did wildlife rehabilitation and the local pythons viewed her small wallaby inmates as a convenient food source—much like when wildlife corridors were installed near one of my field sites up on Cape York. These little bridges were supposed to keep the cute arboreal marsupials from getting squashed on the road like furry cane toads. When the researchers were doing their radio tracking, it didn't take them long to notice that there were marsupials stationary at either end of the corridor, not moving for days on end. Very unusual behavior. Further investigation revealed a large, well-fed python at each end, with a radio transmitter among the blobs of partially digested marsupial in their guts.

The python in next door's yard was striking with its mouth open, revealing the very long teeth used to anchor on to a mammal while it threw the lethal coils around it to suffocate the victim into submission. I knew all too well what sort of tissue damage such dentition could make under the best of circumstances, let alone to the flesh of an elderly person. The snake's variegated body writhed as it moved around erratically, determined to get past this flailing non-predator and to its desired prey. I walked back into my house, grabbed a hessian sack, and managed to get the snake behind the head with my hand in the sack. I then inverted the sack so that the snake's head was on the inside and the long body was sticking out. There was only three feet of snake still sticking out when my moment of glory became a bit less magnificent as the snake everted its bowels all over my face, into my mouth, up my nose, and all down the front of my chest. There is nothing quite like the taste of digested marsupial for breakfast. Remembering a comment by a girlfriend that a couple of shots of vodka was useful for getting the taste of a blowjob out of her mouth, I discovered that four shots will wash away the taste of snake crap. Later that day, a polite knock on my door revealed my neighbor with a freshly baked chocolate cake as a thank-you. So it was all worth it in the end. I'll do anything for chocolate, even go ass to mouth with a python.

My first trip back to the University of Queensland was definitely a massive case of déjà vu, particularly once I found out where I would be setting up my new laboratory. In the eleven years that I had been gone, lots of buildings had changed hands between various schools. Everyone from the Centre for Drug Design and Development was now across campus in the splendidly constructed Institute for Molecular Bioscience, leaving the Gehrmann Laboratories behind. In the intervening years, it had been home to various research groups and, ironically, it was now under the custodianship of my new academic home: the School of Biological Sciences. Not only did I end up in the same building where I did my PhD, but on the same floor and in the same laboratory, with my former supervisor Paul Alewood's office as mine. Needless to say, I found this appealing.

I took a walk across campus and slid into the warm waters of the university pool, where I had spent so much of my time during my PhD years. My first training session felt very flat. Rather than gliding otter-like up and down the lane, I struggled like a waterlogged poodle. I chalked it up to the upheaval and being terribly out of shape. But over the next few weeks, my muscle tone did not improve and I felt surprisingly mentally and physically listless and uninspired. I should have been on a euphoric rush, back at my beloved UQ, making a new life and leaving the smouldering wreckage of my former life behind. But try as I might, I just could not get it up. In any manner whatsoever.

Thus a new problem presented itself, one that had gone unnoticed during the months I was dazed and confused in a hospital while on a variety of opioids and their derivatives. I was unmotivated; my typical energy levels had not returned, even after I'd weaned myself off the horrid painkillers. My wasted muscles did not bounce back in tone or bulk, even when I could muster up the energy to get to the gym. I had no sex drive. My moods fluctuated between black and blacker. My mojo was gone.

Luckily, my Californian ER mate Sean Bush had warned

me to look out for exactly these symptoms. It was something he had noticed in car-wreck victims who had suffered severe spinal injuries resulting in extreme surgery followed by lengthy hospital stays. The combination of the shock to the system and the chemical overload from the months on massive amounts of long-acting opioid painkillers made the body's biochemistry and delicately balanced hormone pathways go haywire. In effect, my endocrine system rebooted full of all sorts of start-up errors. I was no longer Mac; I was now Windows.

After seeing a primary care doctor, I was referred to Dr. Russell Cooper, an endocrinologist who specializes in exactly these sorts of issues. After a battery of different tests, the picture was quite clear. I was producing almost no growth hormone or testosterone, while other elements of the androgen cycle were being made in too large a quantity. The two hormones so very central to being a man were absolutely decimated. For some reason my vitamin D levels were also way down. The amount of sunshine I'd got lately should have more than brought the levels back up from the low they hit during my months indoors, confined to a hospital bed, but somewhere in that synthesis pathway, something was off too. It was collectively a total mess. I had to think back to basic biochemistry courses from my undergraduate days and crack out biochemistry textbooks to re-familiarize myself with these extremely complex feedback pathways. The conclusion I came to was that the extreme alpha-ness that I had taken for granted, and which was such a prominent part of my personal and public identity, was gone.

I did some digging into the scientific literature and was gobsmacked to discover the impact that opioid painkillers had upon testosterone levels. Opioids are classified as short-acting or long-acting. Short-acting opioids release medication quickly and are usually taken every four to six hours. In contrast, long-acting opioids release medication slowly and are generally taken every eight to twelve hours. While long-acting opioids are the most effective pain-management option for people with severe spinal injuries such as my own, and often the only option at all, they are also the ones that affect testosterone levels the most. In one study I came across, 74 percent of the men on long-acting opioids had low testosterone, while for the short-acting opioids the rate was 34 percent. The recovery rate was also markedly different. The risk of the drugs resulting in permanent low testosterone was almost five times higher for men taking long-acting opioids than for those taking the short-acting kind.

Intrigued, I dug further into the literature to find out what other drugs affected testosterone levels, either transitorily or permanently, and was astounded to find that the list was diverse. Antidepressants, antipsychotics, and tranquilizers were among the most potent outside of the opioids, but myriad others also made the list, including those used to treat chronic diseases such as those related to high cholesterol levels (with statins being particularly toxic to testosterone levels), convulsion and epilepsy, and high blood pressure. Furthermore, many recreational drugs had an impact, spanning the range from depressants such as alcohol and marijuana to stimulants such as cocaine. The implications of this are staggering and socially insidious. Low testosterone affects everything from the psychological wellbeing of the man, through to his basic fertility rate, through to being a major contributor to prostate cancer. In short, it was a silent epidemic.

These acute effects are separate from, but still related to, the natural drop in testosterone levels brought about by a man's age. The idea of male menopause is typically blithely dismissed with statements like, "Male menopause is much more fun than female menopause. A female gains weight and gets hot flashes; a male dates younger women and drives a sports car." However, it is a very real medical issue that eventually affects all males to one degree or another. "Andropause" is a much more accurate description of the usual decline in the production of testosterone as a man ages. Obvious symptoms are changes in sexual functioning, such as erections not being maintained or even obtained, or there being no "morning missile," as well as a generalized drop in sexual interest. More subtle signs include changes in emotional stability, with the man becoming moody, irritable, sad, depressed or unmotivated for no apparent reason, or lacking in self-confidence. In my case, it was an extreme variation brought about by physical trauma combined with months on end of massive doses of long-acting opioid painkillers.

The good news is that the diagnosis—which requires blood testing on different days under different conditions—is very straightforward. In my case, the diagnosis was much more complex as I had a number of issues going on, far beyond a simple age-related drop in testosterone levels.

As with everything medical, men are pretty hopeless in seeking assistance, but this is particularly acute for problems in "that" area. This is why prostate cancer in men is far too often not diagnosed until it is too late. Men, including myself, typically try to "man up" about medical problems, gritting our teeth and quietly suffering through them. The perceived social stigma of being seen as "less of a man" exacerbates this. I say "perceived," because I have a strong suspicion that it is only true in our own eyes. This is reflected in the changes in the medical awareness campaigns regarding prostate cancer, for example. After years of unsuccessful propaganda urging men to get checked, now the female partners are being targeted. Women are basically encouraged to nag their man into going to the clinic and getting things checked annually from an early age. Andropause should be treated no differently than menopause, since ignorance and denial can be dangerous.

If a woman has any sort of medical problem, the phone lines ring hot and very intense lunches are scheduled. There are fundraisers, charity runs, public head-shavings, cupcake support days, all organized in the name of raising awareness of, for example, breast cancer. Let me ask you, when was the last time you saw a guy running in an "I survived prostate cancer" T-shirt? By way of comparison, I can count on one crooked hand the number of friends or family who knew what I was going through. Indeed, most didn't have the faintest of glimmers about what was going on, and reading this book will be a revelation to them. I also don't recall being out fishing, impaling a worm on a hook and turning to a mate to say, "So, how's your penis doing nowadays, mate?" Women probably don't start conversations by discussing their breasts either, but after consulting with female friends, women tend to be much more in tune with their bodies, and they disseminate information to each other much better. About the only time men even mention the subject is at a funeral, with statements such as, "Gazza died of nut cancer."

Goldilocks me had to get my biochemical "too hot" and "too cold" to some sort of "just right." Getting things balanced took a lot of trial and error, with some pretty spectacular swings of the pendulum along the way. Androgenic storms are easily the heaviest shit I have ever had to deal with. Nothing in all my chemical wasteland history had prepared me for the emotional intensity of it. At times I would experience a fresh form of madness: complete loss of volitional control of my emotions. This was more psychologically intense than the time I freaked out on psychedelic mushrooms after making the bad move of watching the movie *Scarface* while tripping.

Wrapped up in a web of anger, I would hop on to my Honda Rune motorcycle and ride it at suicidal speeds along the twists and turns of Mt. Glorious, with music like "Stillborn" by Black Label Society infecting my ears. This beast was the 1800 cc love child of a power cruiser and a missile. The 110 horsepower of torque accelerated it with such force that I could pull a wheelie effortlessly even in third gear. I would try to lose myself in the tortured scream of the highly tuned engine, not caring if I lived or died: one thousand pounds of chrome and steel being ridden by a madman who was in danger of taking up a new career as a hood ornament.

In order to get some control of my life, I pruned it back savagely. In addition to stopping drinking and taking all other recreational chemicals, I stripped all my thoughts and feelings down to the absolute skeleton. Listening to the Grateful Dead's song "Touch of Gray," I came to the awareness that the only way I was going to survive was to tear it all down and build it back up again. There was too much mental and physical clutter. Too much emotional and biochemical entanglement. I needed to minimize external impacts, too. This did not mean retreating. Rather, I needed to cull any sort of negativity from my life. I needed to rebuild myself, starting from the biochemical ground floor up. Eventually, an acceptable chemical balance was reached, after several very difficult months. The stasis was achieved through the daily administration of prescription creams, pills, and injections that I now have to take every day for the rest of my life, including testosterone, growth hormone, and dehydroepiandrosterone (DHEA), among others. This comes at a significant personal economic toll, since my private health insurance only covers a small proportion of the prescription costs.

This was all a very interesting personal exploration. "Interesting" in the way of the Chinese curse: "May you live in interesting times." For so long, I had defined myself by my alpha maleness. I could go balls-to-the-wall harder than anyone I knew. But now, from a chemical perspective, I was basically castrated. I might as well be ball-less for all the contributions coming from my shrunken testicles, but my endocrinologist had found the antidote to this chemical kryptonite. I thought long and hard about whether I really should care about losing my mojo when I can artificially recreate it. Does my chemistry define me as a man? Am I no longer an alpha? Am I emasculated? Or should I focus on the positive? It's a simple numbers game—I have too little of this or that, so with appropriately calibrated doses of bioidentical compounds, the balance can be brought back to the fore. And if it can be fixed, is this all that matters? I am now not just the Bionic Man, I am also the Biochemical Man.

Once my hormones were stabilized and my life was no longer one giant emotional roller coaster, I could get down to getting my research back on track. At this time, I recruited a dozen students to establish my new laboratory, including Tim Jackson. In the decade since he had provided such invaluable help with the snake venom research in Singapore, he had followed his other great love: music. He became an electric guitarist of some note, but this alternative career universe was derailed by an idiopathic neurological disorder. With this path now closed to him, he returned to science with a vengeance. He undertook his honors research on the various small elapid venoms I had collected during my PhD years and after my return from Singapore. He showed that not only were they as complex and potent as the venoms of their larger cousins, but they were also packed full of very novel toxins with significant potential for drug design and development. He then embarked upon PhD research in my lab; it felt to me like my prodigal son returning home. My intellectual kid brother Eivind Undheim also made a reappearance in my life, undertaking a PhD on centipede venom evolution under my supervision.

My personal focus was the venom of the long-glanded coral snake, as this research had been derailed by my breaking my back. I recruited a new PhD student, Daryl Yang, to undertake this work by splitting his time between my lab and that of my long-time collaborator Wayne Hodgson. The research into the venom revealed that it caused an almost instantaneous paralysis—not a limp paralysis, like that brought about by a death adder's venom, but a spastic paralysis that would make the snake's prey flop around like an epileptic in a nightclub. The neurotoxins cause the muscles to spasm rhythmically, with each contraction more profound than the one before and very little relaxing between contractions as the baseline climbs higher and higher. As the snake prey tired, it would become more rigid. Inhalation would become progressively more difficult as the diaphragm became more and more tightly contracted and difficult to relax.

Suffocation would occur, but from the opposite effect on the muscles than that of the venom of a death adder or a krait, even though they are all elapid snakes. Instead of a morphine-like relaxation to death, it is a tetanic spasm. The most amazing part is that it is the same general class of neurotoxin as that of the classical elapid snake—the one that relaxes things to death. But in this case the toxin was modified to act directly on the nerve channels and turn them on, rather than blocking the message, as is normally the case. It caused all the neurotransmitters to be released at once, making all the muscles flex like the average guy while shirtless at the beach, but to the point that they are immobile. It was one of the most amazing cases of adaptive evolution in snake venom, found in the most highly refined snake. It doesn't get any better than that.

I was also well overdue to get down to work on the sea snake samples collected over the last decade. The first order of business was to collaborate with Kate Sanders from Adelaide University to genetically barcode every specimen we had collected venom from. This was routine for my research, where I made no assumptions about the taxonomical affinity of any of the specimens, particularly highly confusing species of sea snakes. Three types of samples instantly gave anomalous results. As expected, the sea snake with the very rough scales was confirmed as a new species. As our boat captain had been instrumental in its discovery through guiding us to a variety of different sea floor habitats, we decided to recognize his pivotal role in the discovery by giving the snake the scientific name *Hydrophis donaldi*, with the common name of the rough-scaled sea snake.

This was not the only pleasant discovery. Among the samples I collected were those from the egg-specialist marbled sea snake and the beaked sea snake. The results from the marbled sea snake indicated that it was also a new species, but one still closely related to the population found in Southeast Asia. We gave it the scientific name of *Aipysurus mosaicus* and the common name of mosaic sea snake, in recognition of its intricately beautiful pattern. But the beaked sea snake samples came out as very distinct from the beaked sea snakes in Asia, to the point that they were separated by a large number of other species. This was despite these snakes looking virtually identical. It was a remarkable case of convergent evolution—the natural selection process where two unrelated animals evolve to look very similar due to the independent specialisation for a unique ecological niche. A well-known example in snakes is the similarity between the green tree python in Australia and the emerald tree boa in South America. However, this was the first time that such convergent evolution had been documented in a lethal species.

This had immediate implications for the treatment of envenomed patients, since the manufacturer of the sole sea snake–specific antivenom, the Australian company CSL, sourced the venom used in the process not from Australia but from Malaysia, because the snakes were easier to collect there. Assuming they were the same species, they had also assumed the venom was the same and therefore the antivenom would work well against either.

We worked out the chemical profile of these venoms while in parallel testing the effectiveness of the sea snake antivenom against all of our sea snake venom samples, including that of the beaked sea snake from Australia. For the sake of comparison, we also tested the venom of katuali. To our great surprise all the sea snake venoms came back as virtually identical: streamlined and simple. I concluded this was due to all sea snakes being specialists for a single type of prey: fish. This hypothesis was validated when the katuali venom also came back as extremely simple, to the point that it was virtually indistinguishable from that of the sea snakes. As these snakes also feed exclusively on fish, the evolutionary selection pressure had resulted in a chemical convergence. Even more surprisingly, not only did the CSL sea snake antivenom work with virtually equal effectiveness against all the sea snake venoms, but it also neutralized the katuali venom with the same degree of efficacy. This was unprecedented.

I also branched out into leech venom research. European leeches have been used in medicine for hundreds of years to promote bleeding. Some of the uses are based in superstition and thus of limited efficacy, or even counterproductive. However, leeches have been used to bring blood flow back to fingers that have been reattached after accidental amputation through having them feed on the affected digits. Drugs have also been developed from the powerful anticoagulants present in the leech venom. But Australian leeches have remained virtually unstudied. To jump-start this project, I chose the giant aquatic species present in Victoria. These four-inch-long beasts are a rarity among leeches in that they are predatory, not simple parasites like most leeches. They will feed on a freshwater crab, a frog, or even a small mammal and suck so much blood out that the animal dies. In order to stimulate the venom glands for the research, I needed to feed them. Due to animal ethics constraints, I could not let them feed on live vertebrates and sourcing suitable invertebrates proved impossible. So there was only one resource: me.

I placed eighteen of them on my arms and let them attach to my veins. Almost immediately, I could feel each of their slicing jaws cut deep into my skin like circular saws. They then settled down and fed. Over the course of the next hour, each drained me of over 5 cc of blood. After they were done feeding, they dropped off one by one, like drunks out of the pub right before closing, and I returned them to their containers. Each left behind a hole that was less than an inch in diameter and oozed blood for the next three days. My arms were covered with gauze that was continually soaked through. It was like having another Stephens' banded snake bite, only this time the effects were limited to the immediate area around the wounds. It was gross, but I was not in any danger. However, it was not something I planned on doing again! And the future results will hopefully make this disgusting experience worthwhile.

During this period, one of my all-time favorite studies was undertaken, one that showed that "vintage venoms" lose none of their bite. We discovered that venoms stored for up to eighty years remain biologically active. These were the venoms I had come across at the Australian Venom Research Unit in a dusty storage closet a decade ago. Venoms are a time capsule of disappearing biodiversity and hold potential for the discovery of new medicines. The study examined fifty-two venom samples, including rare and historically important venoms. The research showed that properly stored venoms remain scientifically useful for decades and that vintage venom collections may be of continuing value in toxin research.

Venoms and toxins are a rich source of unexplored compounds which could be used in drug discovery and development. Reptile venoms have been used to develop drugs such as Captopril, which is used to treat high blood pressure, and Byetta, which is used to treat diabetes and has off-label effectiveness as an anti-obesity drug. The venoms we studied came from the invaluable collection curated by the late Straun Sutherland, and their value is a testament to his continuing impact upon venom research in Australia, long after his passing. The venoms of different species have extensive variation, so each venom sample is a precious resource which could contain the next wonder drug. Storing these samples correctly is particularly important as many venomous snake species worldwide are declining and fresh venom may be difficult to come by.

Venomous animals worldwide are disappearing due to habitat destruction, persecution through activities such as rattlesnake round-ups, and the impact of feral animals such as cane toads. Many venomous snakes have disappeared from large parts of their range, or become extinct. Collected venoms may one day be used for research, long after the animals themselves have become extinct. For example, as part of this project we studied death adder venoms from locations where the adders have been wiped out by cane toads.

Some of the Australian venoms we studied may be the only samples ever collected from a range of unique island tiger snakes which are now threatened by habitat destruction. We also studied the first coastal taipan venom collected for antivenom research by Kevin Budden in 1950. It was such an honor to work with these samples, due to their immense historical significance. I got chills holding these vials, knowing the tragic backstory. The young man who collected this venom was bitten in the process, but heroically made sure the snake was taken away for research before he went to the hospital, where he died shortly after.

After spending time planning and organizing my new laboratory, including setting up a battery of complex machinery for the experiments, I set about finishing the venom book I had started while in the hospital. Initially, the book had been a simple exercise to pass the time and keep me from losing my mind from boredom, but it had developed into something quite special. It spanned the full range: from evolutionary theory to occupational health and safety issues, to the unique nuances of veterinary care of venomous animals, to clinical treatment of the envenomed patient. It was the most in-depth book of its kind. I had engaged nearly a hundred co-authors, specialists in all the topics the book covered; it was the greatest team effort I had ever been involved in, and I was proud of how well we had all worked together. The final product was a testament to the dedication and passion of all involved. I gave it the title of *Venomous Reptiles and Their Toxins: Evolution, Pathophysiology and Biodiscovery* and Oxford University Press agreed to publish it. I viewed it as a true milestone in my scientific legacy.

14
MR. AND MRS. SMITH

While my second life at UQ was back on track and moving along, receiving news that my mate Luke Yeomans had been killed by one of the king cobras in his sanctuary in the United Kingdom was like being struck by lightning. He had been working with a particularly large male, using the same close proximity, high-contact approach that had given me chills down my spine when I last visited him. The inevitable had happened: he had been bitten. Despite the predictability, it was still a donkey punch to my head. His death was global news. I was not able to fly over for the funeral, so I said my goodbyes alone while taking a long walk in the rainforest of Mt. Glorious. It was very hard to deal with having lost yet another friend to snakebite, particularly so soon after having come to grips with my own mortality.

On July 22, 2011, I also learned of the atrocity that had occurred in my Norwegian homeland. Anders Behring Breivik, a member of a Christianity-infused, extreme far-right, white nationalist sect, had bombed government buildings in Oslo, killing eight people. He then gunned down sixty-nine more people, mostly children, at a fjord island camp, which was similar to the camps I had enjoyed during my childhood summers spent in Norway. I was deeply shocked and distressed. This incident really drove home to me that religious fundamentalist fuckwits come in all flavors. Not just chocolate or caramel but vanilla as well.

This all weighed heavily on my mind as I started to prepare for trips to collect Arctic venomous octopus species for my new fellowship research. I was planning on traveling across the northern hemisphere for a few months. However, shortly before I was to leave, I received the most intriguing email of my life.

Dear Bryan,

By now I have read several of your articles, then a bit of your old blog and forum. The interest arose after I had a chance encounter on the ARTE channel of a documentary about you working with taipans. The start with the close up of your back tattoos caught my eye. My conclusion is that we should have dinner. Trivial, yet the conversation can just be too interesting (and potentially amusing) to refuse.

I will try to be time efficient here, as several things should sound familiar to you.

—I grew up on military bases, mainly in the Middle East, in a mixed-marriage family. As almost any officer's child, I have rarely lived in one country for more than three to four years. Travel was a constant.

—Addicted to adrenaline, yet safety-obsessed freak.

—I was in biochemistry branch in high school and college, but then transited to philosophy and theology. From there to sociology, and ended up in political science, where I am actually going to do a doctoral degree.

I feel like I should stop here for now. We are actually as different as similar people can be, based on what I've read. I am in fact truly curious about how/if you get philosophy, biology, and sociology to co-exist peacefully in your mind. The evening can be rich in productive (dis) agreements.

Finally, remembering that you don't know me, and that aesthetics is in many cases no less important than ethics: I am, as a matter of fact, very pleasant to look at.

Let me know if you are in Europe and have a slot.

Best,

Kristina

Omnis enim qui petit accipit et qui quaerit invenit et pulsanti aperietur.

In planning my upcoming travels, I had intended to be in Europe for a couple of months. This could end up being interesting. But first, I traveled far up into the Canadian Arctic and hopped onto a research vessel to net octopuses. This letter gave me much to ponder as we headed out to sea.

The vessel was a boat straight out of the most stereotypical fishing-town novel, with politically incorrect crew smoking indoors and telling dirty jokes to pass the time. This rusting heap of scrap metal had water leaking into its lower levels. Not a pretty little member of an overpriced yacht club, but functional in that ten-dollar hooker sort of way. The sleeping arrangements were equally primitive: a narrow bunk with only a threadbare blanket to cover myself with, and a packing cell full of clothes for a pillow. The weather had apparently turned at about the same time I landed. The local spirits must have sensed a Viking was in the midst of the mist and the ocean greeted us with extensive chain lightning—always a lovely sight to behold when in the only metallic object out on the water. The dark blue waves were as erratic as the bathtub of an epileptic giant. The captain reported that there was extensive flooding on the mainland and the airport I had arrived at only hours previously was now closed due to lightning strikes.

I was tagging along on a fisheries department survey of the health of the scallop beds offshore of Canada's arctic coast. Other than liking them lightly seared and served with chilli and lime, I was supremely uninterested in scallops. What I was interested in, however, was the Arctic octopus, a small denizen of the deep whose venom I wanted to compare to that of the octopuses I had collected five years prior in Antarctica. We got straight into the benthic trawling, despite the elements putting on quite the show. Shovelling out sorting trays by hand, trying not to miss anything, was the ultimate arm workout, one that would give a male of any age the kind of forearm normally reserved for serial masturbating teenage boys. There were lots of empty scallop shells, but not a lot of live ones. There were, however, metric tons of invasive lemon-weed coming up in the nets—a sure sign the marine ecosystem was extremely unhealthy. The octopuses were as scarce as an honest politician. An entire day's work turned up only one of each sex, despite the nets being put down for benthic trawling twenty times in the 980-foot depths. They were tiny things, mantles less than one inch across, even as adults. They were, nonetheless, a cool red color.

The next day was calmer and we had whales cruising right by the boat, which got me thinking about being in Antarctica, gazing long at the whales while the whales gazed at us. Different ocean, different whales, different octopuses, different me. But the familiar steel chain clanged and clattered with a percussive, pervasive racket that vibrationally filled the entire ship. Over and over again, the nets brought up only rubble and lemonweed; hardly any scallops and no octopuses. We were in prime habitat for them, and the fishermen normally saw lots. Of course, they also used to see lots of scallops, but rampant overfishing had ruined the ecosystem.

The entire day yielded only two more octopuses, at least fourteen short of what I needed to collect in order to accomplish the research. We even kept working through another storm as lightning seared the sky above and horizontal rain lashed the deck. The next day we moved further offshore and struck pay dirt, landing eighteen octopuses in total. I had that blissful feeling of knowing I had reached the tipping point in accumulating enough samples to run this aspect of the project. On the way back, the bow spray reflected the sunlight like a million airborne diamonds. After a long voyage back, the boat dropped me off at Digby, where I arrived just in time for the annual scallop festival—the one time of year I could compliment a woman by saying, "That's a mighty nice scallop you have," and not get slapped. The following morning I drove down to Halifax to fly to Geneva to meet this intriguing woman who had tracked me down.

Landing in the Geneva Airport with spine intact after a comfortable Virgin Atlantic business class flight, I strode out into the arrivals hall with Rob Zombie's "Living Dead Girl" playing on the iPod. Waiting for me was an absolute vision. Long brown hair. Half Russian, half Gagauz. Dark amber eyes. Very tall. Very elegant. Very beautiful. Sliding into her white BMW M3, we headed out into this city I knew so well and had such fond memories of. I was now, however, seeing it through new eyes. The late summer sunshine kissed Lake Geneva as we hopped into a paddleboat and made our way out into the champagne-tinged waters. After paddling around for a few hours and having a couple of swims in the warm summer waters, we came back onshore for dinner at a splendid Ethiopian restaurant that Kristina knew of, but that I had never been to in my previous stays in Geneva. It was then back out on to the water. The lights of the cigarette racing boat she had conjured up pierced the night as we headed in the general direction of Montreux. This long first day and night together provided ample opportunity for us to get acquainted.

She had rather undersold her background and resourcefulness. She had lived on her own since she was sixteen, when she moved to Moscow to work as a model in order to pay her way through Moscow University, majoring in philosophy. Finding modeling incredibly boring, she quit it a few years later, once she earned an academic scholarship from the University of Geneva to do her master's degree in political science. She had been in Geneva for the last few years. She was, as it turned out, also the darling daughter of the retired Soviet army officer Mihail Formuzal. Her mother is a very blonde, fair-colored Russian from Orenburg, while her father is Gagauzian (from a small region that speaks Turkish but practices Christianity). In the Soviet army, he was not only a hard-nosed soldier but also a body building champion. At the beginning of the Soviet invasion of Afghanistan, he was in the artillery, where it was discovered he had a particular talent for the use of ballistic missiles. Kristina was born in nearby Uzbekistan and spent the first part of her childhood there.

Growing up, her father's soldiers were her babysitters, so she was taught how to fight from a very young age. Either go on a long march in the Uzbekistan desert, or let the colonel's darling daughter do whatever she wanted. Her nickname, apparently, was "The Little General." She became a crack shot, not surprisingly having a particular fondness for long-range rifles, being Daddy's little tomboy after all. She described to me in detail the difference between getting a Russian tank to move and turn—apparently very different processes were involved—and how an American equivalent would differ making the same maneuver; how the controls differed and how the team of two or more moved different levers or pedals to get these awesome machines to do whatever they wanted. She used the same matter-of-fact tone in which another woman might have described to me how Gonoouka shoes differed from Gnukanuka shoes. Righto.

She told me an intriguing story about a little boy in her Uzbekistan neighborhood being killed by a large Caspian monitor lizard. It had bitten him and chewed on him for a prolonged period of time, over twenty minutes. The three-year-old died a few days later. This was the first death I had ever heard of caused by these lizards. Ironically, a report was published in the scientific literature a few months after this conversation between us, describing another venom-induced death from a monitor lizard bite, this time in Nepal. The bite was witnessed by a number of people in the village, including a local wildlife officer. The large Bengal monitor lizard hung on and gave the woman's leg a sustained chew. Again, enough time for the delivery of whatever volume of liquid was currently in the ample lumens of the mandibular venom gland. She died three days later from cardiac failure, after suffering through kidney failure subsequent to the onset of severe symptoms nine hours after the bite. There was no accurate record about the cause of death to the child in Kristina's village but the two species of monitor lizard from the respective regions are very closely related and it would not be unreasonable to speculate that there was a chance the child in her village died in the same way.

She told me more about her fascinating family history. Once the Soviet Union collapsed, her father seamlessly transitioned into politics and was now the governor of Gagauzia. Not to say that all was butterflies and unicorns—quite the contrary, in fact. Eastern

European politics is a dog-eat-dog world, one that even involves physical altercations in parliament. This was exacerbated for a person such as her father, who was striving to inject some honesty into Moldovan politics. In addition to bugs in their houses, he had endured physical attacks, and even attempted assassinations. This included a particularly unsavory one where he was poisoned at an official dinner and was rushed to the hospital for medical treatment.

Kristina knew six languages, and was extremely well trained in the use of martial arts and weapons. She was as much of an adrenaline-junkie, Type-A personality as I was, and had spent nearly as much time in the hospital. We had a hilarious conversation comparing broken bones. We agreed that shattered ribs hurt like hell whenever the ribs moved, which was, of course, any time a breath was taken! I had her beaten twenty-three to ten on total bones broken but she had fractured her skull, which I hadn't. It happened when she was five and tried to learn how to fly by throwing herself off her grandmother's table, flying headfirst into the robust old TV set. Her worst accident happened when she was fourteen while reading, sitting on her windowsill. She was so into her book that she lost her balance and fell two stories, breaking numerous bones. She was also continually getting immunized against everything, however obscure. The constant hammering of the immune system produced in her a weird and unique sort of high. Getting addicted to immunizations was a funny new one to me.

The only distractions during this ethereal time were the forty messages that landed in my phone upon my arrival in Geneva. A biblical plague of crickets had appeared in the building my laboratory was in, invading that whole level. One of my students, who shall remain nameless since he is buried in an outback Northern Territory swamp, had left an untied bag containing two thousand crickets on the floor. Luckily, everyone on the floor was a biology researcher, so they took it in reasonably good spirits and just cracked out the vacuum cleaners. Apparently, a new sport was made of it. I just smiled wryly and put it out of my brain.

The next day we hopped on a flight to Russia. Moscow proved to be an alternative universe. The amusement began for me upon arrival as we were cutting over to where Kristina's best girlfriend, Inna, had parked her Range Rover. The English translation sign said, "Welcome to Terminal 7!" Considering the emotionally dead face of the average Russian, this sign was very funny. Moscow central was best done on foot or by taxi, not by personal vehicles, because there were people like Inna behind the wheel. This tall blonde was the most lethal thing out there in heels. She drove like she was competing in a blood sport. Quentin Tarantino won't have done a driving scene until he does one in Moscow. Just the drive from the airport to the city could be a scene. People were passing us driving on the sidewalk. In front of a school. At 3 p.m. At forty miles an hour. We saw eight accidents on the way to the city.

Russia is guided by the twin forces of inefficiency and corruption. Never before have I seen a culture that expends so much time and energy to get out of doing something that would take a fraction of the time and energy to just do! Absolutely exasperating. Or it would be if I actually had to deal with it. Kristina just handled it all while I worked on my dead face, which was easy, since I had no idea what was being said. But no one knew that. I just had to be uncommunicative in that special sort of Russian way. Relax all muscles in the face and keep the eyes alert, but at half-mast. Not Slavically surly, just deadpan, playing to the well-deserved stereotype of a Russian male. As time went on, I quite enjoyed putting on this anonymity cloak and people-watching. It was a circular black hole of the watcher being watched by the one they are watching.

The reigning oligarchs had poured in oil money at such an open tap rate that despite the low currency exchange there was an amazing array of high-end services and products. The first few days spent hanging out in Moscow were certainly of a standard exceeding that of Oslo. The real tragedy for me was the music. The best (worst) of the eighties on loop. Those songs were painful enough the first time around. Wang Chung should be called Wang Chunder, since that band is the definition of auditory illness.

Despite the music, aspects of Moscow were stunning. We dined in the rooftop restaurant of a thirty-story building, gazing over the city. The lights of the cars driving chaotically below looked like fireflies on crystal meth. I thought of my Norwegian grandparents walking through these same streets. They were posted here as part of Norway's diplomatic contingent at the height of the Cold War. As expected, they were followed by the equivalent of the modern-day Main Intelligence Directorate. My grandparents treated their followers as their own private security service, sending them liquor and nice food while eating out every night. They actually became friends over time. Very handy, too, since the locals could see who was conspicuously behind them and they were thus spared some potential indignities.

My parallel universe with my Norwegian grandparents continued when Kristina and I dined at the very famous Pushkin restaurant, just a few blocks away from the mighty Kremlin. The ambience was new to me: red velvet curtains, old wood, clientele wearing clothes more expensive than my car, impeccably polite waiters, and caviar with champagne made me feel like a Crocodile Dundee accidentally dropped into *Anna Karenina*.

We then walked along the Moscow River at night. Shortly after we passed the Cathedral of Christ the Saviour, a massive statue came into view. It was as big as it was hideous. As I later found out, this is the tallest monument on earth. Taller than the Statue of Liberty or the statue of Christ in Rio. At 320 feet high, it was not proportional, and depicted an enormous—probably 70 percent of the whole monument—man standing on a tiny boat. I imagine it was intended as the display of might by whoever the hero was, but instead it looked like a grown man standing in a child's paddleboat. How and why was this installed in such a central location in Moscow? Surely there must be a story. And there was.

Once I knew the background, it became even more amusing. Turns out there is a regime-favorite sculptor in Russia. He populated Moscow with his works—some good, some questionable. This particular one was a special case. It is now obscured with myths and anecdotes, but the essence of the story seems to be as follows: in 1991–92, the world was celebrating the 500th anniversary of Columbus's discovery of America. The artist decided to create a sculpture of Columbus on his ship, triumphantly gazing at the land in front of him. Ironic, since Columbus had no idea he had discovered a new continent! The attempts to sell the statue to the United States, Spain, and even the countries of Latin America fell flat. The artist then tried to present it to the United States as a gift, hoping to leave the same mark on the face of the country as the French did with the Statue of Liberty. Several times US officials politely refused to accept this awkward mass of metal. When it became clear that this Columbus was not sailing anywhere, the artist got upset and the admirers of his art high up in Russian politics decided to cheer him up. That's how the monument for Peter the Great, one of Russia's most progressive czars, was born. And this is how you offend three nations with just one sculpture.

Peter the Great was a six-foot-tall athletic guy of Eastern European appearance. He dressed progressively for his time (late seventeenth century) and disliked beards. Christopher Columbus was a middle-sized Italian with a belly, who lived two centuries earlier and dressed in plush, pumped-up pants. The transformation was simple: decapitate Columbus and replace his head with one resembling Peter the Great. Why Peter the Great is holding a golden paper roll in his hand, as if he is about to grandly announce something, is unknown to this day. The new statue was ready quickly and was planned for installation in Saint Petersburg—Peter's darling. Peter the Great disliked Moscow and founded the new city—Saint Petersburg—much closer to the Europe he adored; it closely resembled Amsterdam in architecture. The only problem was that the mayor (and the population) of Saint Petersburg also kindly rejected the gift. They rightfully pointed out that a 330-foot-tall statue would overshadow all the majestic old buildings and palaces of Russia's most beautiful city. They also already have a perfectly appropriate statue of Peter on a horse, where both Peter and the horse are life-size. After long and hard thinking, the Italian Peter was put in the middle of Moscow, on a tiny island on the river. Where he is sailing to from Moscow River is unclear, especially considering that he loathed old Moscow and deprived it of its capital city status. Needless to say, Moskovites are not big fans of the monument and only shake their heads when they have to explain to their visitors who the guy on a boat is. Kristina explained all this to me during the night-long walk through Gorky Park, as the iconic Scorpions song "Wind of Change" echoed through my brain.

But far and away the most interesting phenomenon for me was the "Moscow dogs." These are a subpopulation of abandoned dogs that have not only become established and self-sustaining, but display a pattern of behavior that is quite astounding. They use the subway system to travel in a coordinated manner, getting off with deliberation at certain stops during particular periods of the day. They use this both to search for food and also to patrol an unusually extended home range. Intriguingly, these highly socialized and organized packs actually keep the numbers of "ordinary" feral dogs down. While watching them one day, Kristina recounted to me an event from several years prior. It was peak winter in Moscow, with temperatures plummeting to negative four degrees, without even taking into account the "lazy wind"—the kind that does not go around a body but straight through it, chilling even the marrows of the bones. She had just left McDonald's and had half a hamburger in her hand when she saw a starving Moscow dog. She offered the hamburger to the dog, but it gave one cursory sniff, snorted dismissively, and trotted on its way in search of nourishment. She was blown away: if this starving creature refused a McDonald's hamburger, there must be something very wrong with this "meal."

As I was strapping myself into my seat for an internal Russian flight with Kristina, I read in the English language paper I'd picked up in the airport about the Russian hockey team that had just been killed in a plane crash. The article helpfully provided details about several other recent wrecks, all from Aeroflot—the same decrepit airline we were flying on. After take-off, they served a food-like substance. As I looked in wonder at the luridly orange carrots, which I instantly nicknamed Chernobyl carrots, Kristina told the burly stewardess that she was a vegetarian. The stewardess instantly transformed her tray into a vegetarian one by removing the container containing the meat and moved on down the aisle, leaving Kristina with no replacement—just the salad, a cracker, and dessert to enjoy. Classic Russian logic.

Our destination was Lake Baikal, the deepest lake in the world and containing up to 25 percent of the world's reserves of fresh water. It is located just outside the industrial hell that is the city of Irkutsk. Before heading out to the lake, we went deep into the beehive maze of a geological building to unexpectedly drop in on a Russian Academy of Science researcher who works on the ecology of the Lake Baikal water shrew, a small venomous mammal I wished to study. With me keeping a deadpan face in a Russian sort of way, we were able to sneak me deep into a government building with restricted access for foreigners. This was a scientific institute that would have otherwise been impenetrable to me without an extreme administrative burden—a step we simply bypassed. After liaising with a local researcher and organizing the research collaboration, we drove out of the city and into the sanctuary of the evergreen tree forest. After an hour's drive, we were at the shore of Lake Baikal.

The emerald green waters stretched out to the horizon, with the far shore hidden by the curvature of the earth. In addition to venomous water shrews, Lake Baikal was home to three unique animals, found only in this lake but usually living in the ocean. Two of those I particularly wanted to see, and one I wanted to eat. The first two were freshwater seals and freshwater amphipods. The seals were chunky little things, about half the size of the little fur seals I enjoyed watching cavorting in Antarctica. They were extremely bulky for their short length. Being so small, they had an unfavorable surface-to-volume ratio and thus would be sensitive to heat loss. They were shaped like little balloons.

The amphipods were also much smaller than the ones I had collected in Antarctica. They were lurking ghost-like on the murky bottom as I floated above in an ungainly manner. I absolutely loathe drysuit diving, due to the loss of agility. But the water was far too cold for a wetsuit, as it was currently sitting at forty-four degrees despite it being late summer. Kristina went in with a 7mm wetsuit and got extremely cold very fast. Once we had dried off, it was time to hit a restaurant and track down the third animal: the omul, a white fish found only in this lake. It is considered something of a delicacy and I agreed from the first mouthful that it was astoundingly tasty. The only downer was an adjacent business that had as one of their attractions an anemic brown bear crowded into a cage far too small for its great frame. This disturbed me greatly.

After lunch, we went on the weirdest, craziest zip-line run by two young guys who were drinking beer and "yahooing" on it when we rolled up. They had constructed it themselves and likely were drunk at all stages of the building process. They gave us helpful tips like, "Lift your feet so you don't hit that big rock on this one," and "You'll finish this one going very fast since we got the angle wrong, so land feet forward to brace yourself." My personal favorite incongruous comment came at the last stage, a very long run across a valley, where they said, "Don't worry about that tree near the end. It's only an optical illusion that you'll hit it. As long as you're not spinning, you'll be fine." Of course, this was the run where I went into a flat spin and had to stop it by putting my arms out wide to brake. These were not the polite, 100 percent safe zip-lines of Switzerland. This was much more fun!

Kristina had an early introduction to my family when, at 3 a.m. in a Vladivostok nightclub, I happened to glance up at the sixteen-foot screen to see what I thought was my cousin Haakon's face flash across it. I told Kristina, and we watched the entire twenty-minute loop. It was Red Bull filming in Australia, footage ranging from surfing to the nightlife. While at university in Melbourne, Haakon had spent far more time in nightclubs than in the library, so it was hardly surprising that random footage would include him. And sure enough, it was him. He was putting his face right into the camera and making the classic two-finger, devil's-horns, rock party signal. Obviously he had listened when, years prior, I had said, "Come to the dark side, my young cousin. We have brownies."

After several weeks of traveling across Siberia, it was time to return to Moscow to attend Kristina's former-intelligence-officer cousin's wedding (name withheld upon request). This was a fascinating event, full of scary but cool people, including my first meeting with Kristina's father, who was easily the most formidable person I have ever met in my adventurous life. The following day Kristina and I decamped with the bride, groom, and their friends to dachas deep in the woods. One night, while playing various strategy-based board games in between sessions in the sauna, Kristina and I were huddled in discussion about our next move when one of her cousin's colleagues started laughing and said, "Remember . . . we are spies. We can actually hear you better when you whisper!" This was followed up in the next sauna session when the intelligence officers had a quick conversation in cryptic Russian after I told them all about my venom research and where it had taken me. One of them announced, "We have decided we like you and we would do anything to help you at any time." To which I replied in a smart-ass way, "Even help me hide a body?" I wasn't sure if he was joking or not when the hairy guy from Armenia gave a huge grin and said, "That is my specialty!"

After such a mind-blowing trip, it was time for us to head back to Geneva. It had been late summer when we departed Geneva, but now it was well into autumn. The approaching winter was heralded by a heavy snowfall. I hadn't seen snow since Antarctica years before, so I immensely enjoyed the snowball fights with Kristina, who has a bullet-accurate right arm that was testament to her military upbringing. While she was the most capable woman I had ever met, she was also the worst cook on the planet. This was brought into particularly sharp relief one night when I was preparing dinner and gave her a seemingly unscrewable-up task: to wash the salad. Twenty minutes later she presented me with a bowl of what looked like seaweed washed up onshore after a violent storm. Once I stopped laughing and could speak, I asked her what she'd done to it. She said she'd washed it. As in running hot water on it full blast for about ten minutes, followed by vigorous rubbing between her hands. It was washed, all right!

We spent the next month traveling around Western Europe, including Kristina accompanying me to Oxford University, where I was giving a seminar. After that, we headed over to the reptile expo in Houten, Holland. She was able to meet those of my European friends who were in attendance, including Iwan Hendrikx. She was fascinated and intrigued by this rogues' gallery and also smitten with some of the animals in the collection Iwan and I share at his residence in Holland. The mambas the color of high-quality emeralds in particular drew her attention—one beautiful and lethal creature being drawn to another.

It was time for me to return to Australia without her, but with plans to meet up again as soon as possible. I was scheduled to do another documentary on Komodo dragons six weeks later, so we arranged for her to join me on set and on-screen. I was taking my princess to meet my dragons. The expedition doctor turned out to be the one from my emergency surgery during the Komodo dragon filming trip with Kevin Grevioux. She took one look at me and said, "You won't remember me because you were unconscious, but I removed a pebble from your rotting knee!" I certainly didn't remember her, but I liked her already, and it wasn't long before her services were needed again.

Things quickly went pear-shaped in Flores when an ice cream Kristina was eating was found to have broken glass in it. She had felt several little bits while eating it but thought they were just pieces of ice, so swallowed them. A shard of glass could open her intestines like a zipper, resulting in a painful death. I knew all too well, after my knee debacle several years prior, that the medical facilities on Flores were primitive at best, and the nearest equipped hospital was in Singapore, six hours away by plane. We had two options: to cancel the shoot and make a run for Bali, or to continue on. The first flight out was not until midday the next day—at which time she would either be fine or dead. In order to increase the odds of the favorable outcome, I had the idea of Kristina eating massive amounts of dried mango. The sticky fibres would trap any object, even something as sharp and deadly as glass. To everyone's great relief, it worked and she passed the glass out of her system without incurring an injury. My kind of woman—one so tough she can eat glass and survive! My great relief was followed by deep contemplation about her and our respective futures, something I had already been deep in thought about before leaving Australia.

The following day, we dropped anchor offshore of Rinca Island, ready to commence filming in the morning. While Kristina and I were at the tip of the bow, watching the fiery red sunset reflect off the water and bleed all over the beach, I held her in my arms, looked deep into her amber eyes, asked her to marry me, and slid onto her finger a ring with an emerald centerpiece and diamond satellites. She replied with an enthusiastic "Da!" Love is the ultimate drug. The chemical reaction of falling in love is the most exquisite high that there is. Nothing can top it. I could not believe how quickly the tide had turned and life had bloomed anew. And so concluded the ultimate Hollywood cliché moment of my life.

The last time I had been filming Komodo dragons, I was with the BBC and was a shambling wreck due to my undiagnosed broken back and associated rampant consumption of opioids and benzodiazepams. This time, however, I was clean and healthy.

My feet did not even touch the ground. It was the cruisiest, least stressful, most enjoyable film shoot I had ever been on. Great crew, and freshly engaged to be married. With Kristina's own unconventional upbringing and her past animal experience, she was adept at working with the venomous animals despite not having any experience with these particular creatures. She was also extremely protective of her "Zaya," as she had taken to calling me. I almost burst out laughing one time at the death stares she was giving a Russell's viper one night after we caught it in a ranger's hut. If looks could kill, that snake's mortal existence would have been snuffed out like a candle flame in a cyclone. The only minor mishap was when I was filming a dive sequence and my bare chest scraped across a big patch of fire coral, which quickly lived up to its name.

After the film shoot was over, we flew back to Geneva to pick out the wedding rings. We took the train to Zurich to go to Tiffany's, where we selected simple platinum bands. By this point, winter was well upon Europe, so we decided to travel up to Norway to spend Christmas skiing with my cousins Haakon and Wilhelm and my aunt Vivian and uncle Hans at the mountain cabin, and then to join Kristina's parents in Prague for New Year's. My family, of course, took delight in getting to know her and I discovered her parents were welcoming and warm. Despite his fearsome background, her father turned out to be a thoroughly decent person.

After that, Kristina came to Australia. Upon arrival, she took one look at my unconventional home decorating and announced that it would all be changed. She immediately took charge. This was also when I was able to experience the legendary possessiveness of a Russian woman. "Has this been touched by a whore?" was her subtle way of asking if something in my house had been used by, or given to me by, any woman previously. She played "Possession" by Sarah McLachlan on the iTunes, remarked coldly, "That sounds like whore music," and then dragged all the songs by that artist into the trash. To be fair, it was music contaminating my playlist that had been left behind by one ex-girlfriend or another. It was pretty obvious it was not my choice—Kristina was familiar enough with my musical proclivities to know that my tastes ran to a steady diet of heavy metal. The steel shelves were moved out of the rooms and the rooms used for living rather than as gear depots. I could tell that all I would ultimately be left with would be my books and a few select items of clothing.

Watching all this unfold with a wry smile, Tim Jackson said that my house had definitely needed a woman's touch. This was fine by me, since I had zero interest in domestic affairs—a fact made evident by the thick layer of dust over everything, which Kristina was now cleaning off with the kind of industrial application of potent disinfectant usually reserved for an Ebola outbreak hot zone. "When was this last cleaned?!" she asked. I thought back to how long it had been since I moved in and almost answered with the exact number of months (in two digits) before wisely realizing this was a rhetorical question.

During this cleanout and organization of the house, she noticed that I had a recently received a Christmas card from an ex-girlfriend. "Why is Vampirella sending you a card?" she interrogated. Even though I am a pretty clueless guy when it comes to the workings of the minds of women and rarely pick up on even the most obvious of signs, this one registered. "How did you know her nickname was Vampirella?" I enquired. She gave me the most wicked grin I had ever seen and said, "When I first saw the taipan documentary on you, the one that opened with a close-up of the abstract snake tattoo on your back and then cut to you singing loudly along to 'Highway to Hell,' I was extremely interested. But before I committed myself to anything, I obtained more information." It turned out that she had pulled some strings and hacked my email accounts with the same effort it takes to open a letter, while her equally trained-up girlfriends Facebook-stalked me. All of this was compiled into a proper dossier that was named—I kid you not—Operation Jungleboy. While I should have been scared, I was instead even more intrigued by this amazing woman. Basically, I had met a Russian Bond girl. Or, should I say, I had been selected by one, since women like this aren't picked up. They choose. Kristina was far and away the most dangerous creature I had ever encountered, and I was enthralled.

We went out for road cruises around Mt. Glorious, encountering many snakes each night. Kristina took great delight in moving big carpet pythons off the road with supreme confidence. She spurned the use of a snake hook, instead enjoying the feel of the muscular bodies as she used just her hands. This changed one day when a very small baby python turned and bit her on the finger. She was shocked to discover that carpet pythons are usually quite snappy, and that all the big ones she had been blithely moving off previously were being unusually quiet.

One particularly amusing toxic encounter occurred when she stepped on a stinging nettle. The Australian varieties are particularly potent, causing first blinding pain, and then radiating numbness. She was quite offended. "Why would a plant do this to me?! I love plants. I'm a vegetarian." Once I stopped laughing I pointed out that to plants she was the enemy as they'd evolved this toxic defense specifically against grazing mammals! I added that I didn't get stung because I was a carnivore so the plant did not consider me a threat.

A few weeks later we went up to Weipa with my mate Nick

Casewell, who had come over from the United Kingdom for our fish venom project. Kristina loved catching the stingrays and catfish and was always very put out when she reeled in a shark instead, referring to them as "jerk-faces." During the night-time snake catching, she proved herself a quick study in the capture of sea snakes, including pulling in a new specimen of the rough-scaled sea snake, only the fourteenth ever found and the first one not captured by me.

Next up was starting our little family, with the arrival of two British Staffordshire Bull Terrier puppy brothers—one black with a little bit of white, one white with a little bit of black, in moo-cow small polka dots. We named them Salt and Pepper. Kristina, having only had military-trained Dobermans, didn't even know what a Staffy was, let alone how weird they can be. As Kristina was planning on renovating anyway, it was of no great concern when they started chewing on the cupboards. I put chilli paste on the outside of the cupboard doors. They gave me looks of "Thanks for the sauce, Dad!" as they tore into the cupboards. We came home to them one day, not long after they were neutered. They still had the cones on their heads, but that had not stopped them from destroying the cupboards, drywall, and everything else they could sink their teeth into. Everything and everywhere those small but powerful jaws could reach, they had attacked. I took a photo, gave it the caption "Resistance is futile. You will be assimilated. Love, Salt & Pepper," and posted it on the Dogshaming site. It went viral. Friends of mine come across it on totally unrelated sites. It even showed up on the Russian search engine yandex.ru.

Kristina finally came to the inevitable, evidence-based, rational conclusion: these were aliens from another planet pretending to be dogs. And getting it so very wrong. They had read the manuals on how to be a dog, but then mixed things up because of their extreme attention deficit disorder. They have no idea what is going on, but they think it is awesome. Salt, the white one, appears to be going through life completely ripped twenty-four hours a day; there is a special sort of chemical imbalance going on in that little airhead. He and I shared a special bond immediately.

Our wedding was to be held in Moldova later that year. To get there I first flew up to Hong Kong to stay with Paolo Martelli from my Singapore days, as he was now a vet at the iconic Ocean Park. This was primarily in order to get a tuxedo for the wedding hand-fitted at the excellent Noble House tailors. However, while there I spent time at the live fish markets getting additional octopus and cuttlefish specimens for that research, and then at night wandered the storm drains catching cobras with Anne Devan-Song as she radio-tracked her transmitter-implanted green tree vipers. I also took advantage of the opportunity to go over to the Kadoorie Farm and Botanic Garden in the New Territories to milk some slow lorises for that venom research.

The day before I was to leave, I was speaking on the phone with Kristina. "Since my connecting flight to Moldova from Istanbul is only an hour, it shouldn't take more than a couple of hours to drive over. Can you come and pick me up instead, rather than me waiting in the airport for thirteen hours, as I'm scheduled to do now?" She gave a very amused snort and replied, "You will never shake certain parts of your American heritage: the love of donuts and geographical unawareness! You obviously have not even looked on a map as to where you are flying. Otherwise you would have noticed the Black Sea in between Turkey and Moldova!" I had nothing to say in reply to this! As always, she had more information than I did.

Landing in Turkey, I checked into the very nice hotel located inside the transfer lounge, caught up on some sleep and emails (though me being the obsessively linked-in me, it was not in that order) and then flew on. I landed in Moldova with Zucchero's "Baila Morena" playing on the iPod and was met by my darling girl.

Once in Moldova, Kristina and I took her much younger brother Egor out snake catching. He had never seen any of the local snakes. Her father's number two driver/bodyguard drove us—a very fit guy in his mid-twenties, fresh out of several years in the army, where apparently he was quite the boxer. We compared stories of our noses being broken in boxing and agreed that the potential hit to the pride exceeds even the pain of the break itself. Disfigurement is never cool!

I suspected vipers would be found in the areas with intact old forest, but that the abundant lakes should hold some sort of harmless water snake. So I had us taken to the latter type of habitat. The first lake we went to was full of life. Black geese and dusky ducks littered the water. We also saw several frogs leap acrobatically into the water—an excellent sign, since such creatures are the preferred prey of water snakes. Our first reptile was not a snake, but something I hadn't even thought of finding here: bright green larcertid lizards. I caught one of these living jewels and Kristina and Egor marvelled over it, never having seen the like. We continued on for another 330 feet before I noticed an erratic zigzag carved out of the grass. Despite not being able to see it, I knew it for what it was: a snake. I jumped forward and got it by the tail, lifting it up for inspection before getting my hand anywhere near the head. The blotch-covered body and orange cheek markings identified it as a water snake. I controlled the head while Egor held the rest for close examination. He then discovered the true joy of water snakes, something I had discovered so very long ago as a child—the cloacal secretions of death. Later on, he said that he had washed his hands three times, but they still stank of water snake shit. Yup, water snakes can be that way. But it was not as bad as a viper bite. Though a neat freak like Egor would actually consider taking the venom over the smell!

All weddings require a spectacle in the form of the big first dance between bride and groom. Kristina had sold me on the idea by saying that learning a dance would give us some relaxing time together in the two weeks leading up to the wedding. At the first meeting with our dance instructors, they asked what style we wanted. With utter naiveté, we said that we wanted to do a waltz, being as we were beautiful and elegant people. They asked if we'd had any prior training, to which we replied "Nyet." But, being a pair of blithe spirits, we thought to ourselves, "How hard could it be?"

In our first lesson, it was acutely apparent that we were quite simply incapable of keeping a beat or following a melody. We danced like a pair of automatons. At one point the instructor said I was meant to be pulling Kristina back to me as if she were a graceful swan, not like I was yanking a crocodile out of a pond! Knowing that we would be having our wedding in front of several hundred guests, including diplomats from various corners of the world, Kristina was now understandably completely freaked out. She went running for the shelter of the bride's little helper: valium.

Every day we practiced to the wedding dance song, "Until" by Sting. Being a pair of intellectuals, we finally gave up on the idea of flowing with the music and instead approached it analytically, memorizing that when *this* sound occurred, we would do *that*. The big lift number caused a few twinges in my back, but I worked through it. Slowly and painfully we ended up making a serviceable job of the wedding dance and felt that we would not be complete embarrassments on the big day. I did, however, develop a loathing for that bloody song. Particularly the opening instrumental part, since we often did not even make it to the vocals before we'd screw up and have to start all over again. If I could put that song in a bottle, I'd happily borrow a sniper rifle from one of my in-laws and put a bullet in it. Or perhaps use an RPG, since Kristina had remarked offhandedly one time that she loved the way they sounded when you fired them.

On the day of the wedding, while Kristina and her brides-maids were getting their hair and make-up done for what seemed like an eternity, I headed out with my mates who had shown up for the wedding. First we went bowling, during which Howie McKinney—who'd been one of the original group of scientists in Berkeley self-experimenting on LSD after it was first synthesized in the 1960s—came out with an absolutely priceless comment: "The only problem with no longer taking acid is that I can't tell the pins to fall over." Despite this communication breakdown, he seemed to effortlessly get strike after strike without, apparently, having touched a ball in decades. While I had told my friends about Kristina's family background, any lingering doubts that may have existed were erased when her cousin displayed a prowess at the shooting range that was like something out of a James Bond movie. There was but a frayed hole in the middle of the target's forehead; he even put shot after shot in the same position while the target was in motion, being moved further and further back. When he first started dating his wife, he took her to a Moscow carnival and she went home with the giant stuffed teddy bear that no one ever gets. Rumor has it that he is still a student in compar-ison to her father, though!

It was then time for the wedding, which took on a spy thriller tone the instant the first bodyguard appeared with a submachine gun concealed within an oversized coat. He was merely one of many. Some of them were wedding-party-specific; others accompanied various Eastern European diplomats who had graced us with their presence. In the midst of this was the ambassador from the United States, William Moser, a refined gentleman with impeccable manners. I had a very congenial conversation with him, and he took great delight as I detailed the Byzantine pathway by which I had ended up in this corner of the world, under such unusual circumstances.

It did not take long for the first diplomatic incident to occur, when the group from Cyprus, one of the first arrivals, sat at the table reserved for the Greek guests. These two countries have a long and acrimonious history, so our wedding became an opportunity to foster the conflict. This caused a complete reshuffling of the seating arrangements, not only for the Greeks, but requiring careful consideration of the surrounding tables too. The seating tags for various allies were moved with chess-like precision around the room in a five-minute frenzy handled with great aplomb by Kristina's lovely mother and the unflappable wedding planner. As an impartial observer and avid people-watcher, I viewed this with fascination and concealed hilarity. It was great spectator sport.

The attendees included a diverse who's who of Eastern European society, ranging from farmers, to military, to presidents, through to some who would not look out of place in the Russian version of *The Godfather*. One particularly scary guy came up to me; he had scars down his face and neck. He said in a very friendly voice, while shaking my hand and with a big smile on his face, "I am so happy for you! Kristina has been great friends with my daughter since they were three. I love her like a daughter. Take good care of her!" He then gave me the classic, chilling Russian dead-face and added in a flat tone that spoke volumes, "Or else." It turns out this was her father's best mate and they had survived the military together. I was under no illusions as to the fate implied by "Or else."

My mates Mickey Bhoite, Chris Clemente, Iwan Hendrikx, Anna Nekaris, and Holger Scheib were there, as were other friends and, of course, my family, including my cousin and great friend Haakon. My father made a particularly priceless joke that perhaps there was a time when he was stationed in Europe that he and Kristina's father had spied on each other. He and her father dissolved into laughter, though this joke was probably not far off the mark!

My best man was Iwan and the two of us were in fundamental agreement that this was easily the most dangerous situation we had ever been in, together or at any time. Pakistan was a walk in the park compared to my own wedding! There arose the philosophical question: how does one impress a room full of trained killers? By showing footage on a large screen of the groom and best man in Malaysia with the king cobras we filmed for the *Asia's Deadliest Snakes* nature documentary. We agreed that the only way not to go insane during a wedding ceremony such as this was to go into it already insane!

Men always want to be a woman's first love, while women like to be a man's last romance. It is motivating to know that if we ever have a marriage breakdown, rather than sleeping on the couch, there is a shallow grave in Siberia with my name on it. This was brought into sharp relief seconds after the marriage vows, while we were still up on the podium. After sliding the ring on to my right index finger in the proper Russian way, Kristina told me that the start of our honeymoon was a gift from her father in the form of VIP center-line seats to the Euro 2012 football finals: Spain versus Italy in Kiev, Ukraine. Close enough to see the grass fly from the golden boot of Fernando Torres as he scored another winning goal.

Also close enough to see Mario Balotelli do his imitation of the dying swan dance in a shameless attempt to earn a penalty kick.

She then leaned over and whispered lovingly in my ear, "Whatever you do, never forget one thing. Being married to me is like being in a submarine at ten thousand feet. There is no way out alive."

ACKNOWLEDGMENTS

Wow. After surviving so many crazy adventures across the globe, there is literally a cast of thousands to thank. First and foremost, my heartfelt gratitude to my parents for putting up with such a lunatic of a son. I know it was very difficult at times! I would like to express my everlasting love for my wife, Kristina, who is quite simply the most amazing person I have ever had the privilege to get to know. You are the best thing that has ever happened to me. Thank you for putting up with the idiosyncrasies that make me a difficult person to live with sometimes! Sorry for leaving the towel on the bathroom floor. Again.

And of course no acknowledgements section would be complete without a shout out to my long-time partners in crime, Chris Hay and Iwan Hendrikx, without whom many of the animals would not have been successfully found and captured for the research, and who ensured I returned alive (if not intact) from far-off lands.

My cousin Haakon Teisner was an absolute legend on several adventures, in addition to being a great friend to me. I am incredibly grateful to my doctors Justin Paquette and Russell Cooper for bringing me back from the edge after I shattered my spine and then had my hormone pathways nuked by the pain medications. I am also forever in the debt of Professor Paul Alewood for accepting me as a PhD student and thus launching my scientific career. Other people who deserve special thanks are Alejandro Alagón, Syed Ali, Steve Backshall, Mickey Bhoite, Naomi Borg, Leslie Boyer, Diane Brandl, Nick Casewell, Myke Clarkson, Christofer Clemente, Chip Cochran, Tom Crutchfield, Lauren Dibben, David Donald, Ed De Grauw, Nathan Dunstan, Kris Foster, Dessi Georgieva, James Haberfield, Alan Henderson, Timothy Jackson, Marc Jaegar,

Rob Jones, Elazar Kochva, Joshua Lyon, Paolo Martelli, Karthi Martelli, Gerry Martin, Nigel Marven, Devon Massyn, Richard Mastenbrook, Sean McCarthy, Howie McKinney, Ernest Minnema, Peter Mirtschin, Ray Morgan, Jarle Mork, Soham Mukherjee, Anna Nekaris, Stuart Parker, David Quigley, Kim Roelants, Stefanie Rog, Zach Serhal, Gowri Shankar, Arun Sharma, Holger Sheib, Cindy Steinle, Wilhelm Teisner, Eivind Undheim, Harold van der Ploeg, Nicolas Vidal, Freek Vonk, Galadriel Walter, John Weigel, Guido Westhoff, Rom Whitaker, Desiree Wong, and Wolfgang Wüster.

Of course I have to give my deepest thanks to the funding agencies that paid for the research expeditions including the Australian Antarctic Division, Australia & Pacific Science Foundation, Australian Research Council, Herman Slade Foundation, Human Frontier Science Program, University of Melbourne, and the University of Queensland. I would also like to thank the various documentary channels for filming the expeditions including Animal Planet, BBC, Discovery Channel, National Geographic TV, and the Smithsonian Channel. I would like to express my appreciation to the Explorers Club for accepting me into their elite ranks; this was truly a childhood dream come true.

If I have forgotten to mention someone or a particular organization, please accept my apologies!